Where Have All the Wildflowers Gone?

Also by Robert H. Mohlenbrock

*Wildflowers of Fields, Roadsides and
Open Habitats of Illinois*

Spring Woodland Wildflowers of Illinois

Forest Trees of Illinois

Illustrated Flora of Illinois (10 Volumes)

Co-Author

*Wildflowers of the United States:
Volume I: The Northeastern States*

*Spring Wildflowers of
Carlyle-Rend-Shelbyville Lakes*

Where Have All the Wildflowers Gone?

A Region-by-Region Guide to Threatened or Endangered U. S. Wildflowers

Robert H. Mohlenbrock

ILLUSTRATIONS BY MARK MOHLENBROCK

MACMILLAN PUBLISHING CO., INC.

NEW YORK

COLLIER MACMILLAN PUBLISHERS

LONDON

9 / 52

Macmillan Publishing Co., Inc.
866 Third Avenue, New York, N.Y. 10022
Collier Macmillan Canada, Inc.

Library of Congress Cataloging in Publication Data
Mohlenbrock, Robert H.
Where have all the wildflowers gone?
Includes index.
1. Rare plants—United States.   2. Wild flowers—United States.
3. Endangered species—United States.
I. Title.
QK86.U6M63   1983   581   82–23411
ISBN 0–02–585450–X

10 9 8 7 6 5 4 3 2 1

Printed in the United States of America

Acknowledgments

Grateful acknowledgment is made to the following people and institutions for permission to reprint the color photographs: Plate Nos. 1, 7, 10, 13, 18, 19, 24, 25, 26, 27, 35/Robert H. Mohlenbrock, Southern Illinois University; 2/Richard S. Mitchell, New York State Museum; 3/Laverne Smith, United States Fish and Wildlife Service; 4, 11/Robert Sutter, North Carolina Department of Agriculture; 5/John Garth, University of Georgia; 6, 8, 9, 12/George Folkerts, Auburn University; 14/J. Kenneth Dean, New York State Museum; 15, 16, 17/Susan Crispin, Michigan Department of Natural Resources; 20/Harold Hungerford, Hungerford and Associates, Carbondale, Illinois; 21/Eric Ulaszek, Darien, Illinois; 22/Bonnie Amos, Baylor University; 23, 29, 31/Larry England, United States Fish and Wildlife Service; 28 (from a cultivated plant at Northern Arizona University), 32, 33, 34/David Mueller, Carbondale, Illinois; 30/Elizabeth Neese, Salt Lake City, Utah; 36/Mary DeDecker, Independence, California; 37/Alice Q. Howard, University of California, Berkeley; 38/Lawrence Heckard, University of California, Berkeley; 39/Ralph Faust, Coeur d'Alene, Idaho; 40/Sheila Conant, University of Hawaii.

To my family,

Beverly, Mark, Wendy, and Trent,

who assisted and encouraged me during every

phase of this project

Contents

Foreword

The ecological revolution of the 1960s resulted in a number of environmental actions and an awakening of many persons in this nation to issues relating to plants and animals (including man) and the world they live in.

One of the positive results was the concern shown by this nation in the protection of our wild plant and animal life. The Endangered Species Act of 1973 went a long way in providing legislation under which plants and animals that are on the verge of extinction can receive maximum protection. The listing of endangered plants began on August 11, 1977, after intensive study was made of plants proposed for the list. By early autumn 1982, sixty-four plants had been cited. Scores of others are under review by the United States Fish and Wildlife Service.

This book attempts to provide an account of each kind of plant that has been placed on the federal list of endangered and threatened species, as well as those that have gone through the listing process except for the final listing. In addition, nearly sixty plants in the United States that are rare and are being reviewed by the United States Fish and Wildlife Service are described in this book.

Following an initial chapter that discusses one of this nation's extinct plants and the efforts now being made to prevent further extinction, there are seven chapters based on geographical regions in this country that include the discussions of the endangered and threatened plants. Those plants which have already been listed by the federal government are given first in each chapter, followed by those currently under review.

Most of the plants covered in this book are herbaceous wildflowers, but a few woody plants, such as the Virginia round-leaf birch, Sonne's barberry, and Raven's manzanita, are also included because they have been listed by the United States Fish and Wildlife Service as federally endangered plants. Although these plants are woody, they do produce flowers, just as the herbs do.

The scientific names used for the plants discussed in this book are those recognized by the United States Fish and Wildlife

Service in their official list of endangered and threatened plants. Since plant taxonomy, which includes the naming of plants, is an ongoing and continuous study, the names of plants can sometimes change quickly. Therefore, some of the plants in this book may have a different name after this book has gone to press.

Many people have assisted me in the preparation of this book, and to each of them I give my most sincere thanks. I appreciate the help of all persons who provided photographic slides, information concerning the various species, and assistance in the field. I acknowledge the curators of the herbaria who loaned me specimens to study and to have illustrated. I extend my gratitude to the Missouri Botanical Garden, whose cooperative staff and excellent facilities enabled me to proceed with my study. My editor at Macmillan, Mrs. Lea Guyer Gordon, has generously contributed her time and expertise to this book. Her assistance is gratefully acknowledged.

I give particular thanks to my family, all of whom accompanied me on many of the field trips, and specifically to my eldest son, Mark, who provided all of the illustrations, and to my wife, Beverly, who typed the many drafts of the manuscript.

Robert H. Mohlenbrock

September 1, 1982

Prologue
The Endangered Species List
(with apologies to Gilbert and Sullivan)

And one day it did happen, as our plants were getting rare,
We made a little list, we made a little list
Of some species that were nearly gone, but of which we gave a care.
They surely would be missed, they surely would be missed.
There's the pretty Furbish lousewort growing up in northern
   Maine,
The San Clemente larkspur living on its flat terrain,
The mountain golden heather growing up on Table Rock,
And milk vetches, phacelias, and MacFarlane's four-o'clock,
And Chapman's rhododendron, which is rare, I must insist.
They surely would be missed, they surely would be missed.
We've got 'em on the list, we've got 'em on the list,
And they surely would be missed, they surely would be missed.

And the plant that's called the silverling and Todsen's pennyroyal,
We've got 'em on the list, we've got 'em on the list.
And the Contra Costa wallflower living in the sandy soil.
It surely would be missed, it surely would be missed.
And there's Raven's manzanita known from just a single plant.
I'd say that that is pretty rare, I'd even say it's scant.
And the cinquefoil from the mountains that is mighty hard to find,
And the Navasota's ladies'-tress and others of its kind,
And the monkshood from the northern states, a plant you can't
   resist.
I'm sure that they'd be missed, I'm sure that they'd be missed.
We've got 'em on the list, we've got 'em on the list,
And they surely would be missed, they surely would be missed.

And the Texas poppy mallow, according to our notes,
It just barely does exist, we've got it on the list.

The San Clemente paintbrush that is plagued by feral goats,
They'd all of 'em be missed, they'd all of 'em be missed,
And all those prickly cacti that are very nearly gone,
Such as Pediocactus, Sclerocactus, likewise, on and on,
And pogogyne and stenogyne and also you-know-who.
The task of filling up the blanks I'd rather leave to you.
But it certainly does matter what we place upon the list,
For they'd all of 'em be missed, they'd all of 'em be missed.
We must put 'em on the list, we must put 'em on the list,
For it's too late when they're missed, it's too late when they're missed.

Where Have All the Wildflowers Gone?

❈ 1 ❈

The Disappearance
of Thismia

It was overcast in Chicago when Norma Pfeiffer awoke that August 1 morning in 1912, not a suitable day for a field trip, but Miss Pfeiffer had been planning for several days to visit a patch of prairie south of the University of Chicago campus to look for material suitable for her research. Little did she know that by the time the sun would finally glide from behind the broken cloud cover at noon that she would make one of the most remarkable plant discoveries in the history of botany.

By midmorning the young botanist, who had turned twenty-three earlier in the year, arrived at her destination, a flat, open prairie rather densely covered with vegetation of varying heights. She slowly made her way through waist-high tussocks of black-eyed Susan, thoroughwort, and various kinds of goldenrods, kneeling occasionally to push aside these coarse herbaceous stems to observe what lay at the surface of the soil. Several low-growing mosses were crowded close together in the shade provided by the goldenrods, and an occasional filmy strand of a small clubmoss, or *Selaginella*, hugged the soil. In damper areas, the handsome blue iris was giving way to the pink of the swamp milkweed.

Suddenly Pfeiffer observed an astonishing sight. Lying between several small mounds of mosses were tiny white to pastel blue-green swellings, appearing to have emerged recently from the soil. The largest of these was about one-fourth of an inch tall above the ground and of about equal size across. As she parted the stems of nearby herbs, Norma saw several more of these nearly transparent structures, some just barely visible above the soil. There were no leaves, and nothing about the specimens was deep green, which would have indicated the presence of chlorophyll.

Hurriedly but carefully she removed the soil from around her discovery and prepared to take some of the material back to the laboratory, where she would attempt to identify it.

The above-ground part of the specimen narrowed abruptly where it entered the soil and merged into nearly colorless, thread-like, horizontal roots. The idea that she had found an unusual transparent moss was dispelled by the unearthing of the root system.

At the laboratory, the excitement grew as first one fellow student and then another each in turn proclaimed his astonishment and bewilderment at this plant. Professors John M. Coulter, Charles J. Chamberlain, and W. J. G. Land, considered among the most respected botanists of the country at that time, were puzzled by the find.

Under the microscope the blue-green transparent mystery was observed to be a flower! The flower was tubular-shaped, with a raised ring around the top. Hanging from this ring to the inside of the tube were six tiny pollen-producing stamens. Above the opening of the ring, the blue-green petals were prolonged to form an arch.

The flower was nothing at all like any other flower known in the Chicago area. The total absence of the green pigment chlorophyll was also mystifying, because only a relatively few flowering plants are nongreen.

A search of the botanical literature in the University of Chicago library finally revealed that the little plant from the prairie south of the campus belonged to a tropical family of plants known as the Burmanniaceae, a group closely related to orchids. More precisely, the new plant was some type of *Thismia*. Of the fifteen kinds of *Thismia* known in the world, none had ever been found before in North America, the nearest being several thousand miles from Chicago. What's more, all fifteen previously known Thismias grew in rich-loamed primeval forests, in regions of great rainfall.

Having made such a remarkable discovery, Norma Pfeiffer and her professors made several trips back to the prairie for additional observations. By mid-September, some of the little flowers had developed into tiny fruits with minute seeds.

Throughout the winter of 1912, Pfeiffer studied every aspect of her specimens. On July 1, 1913, she found more of this new plant in the same prairie. She surmised that the underground parts had overwintered. Her studies indicated that she had found a new species, which she named *Thismia americana* in 1914.

In August 1914 Pfeiffer again observed *Thismia americana* in the prairie south of the campus. Little did she know that this would be the last time it was ever to be seen alive.

Botanists from all over the country have tried to relocate *Thismia americana,* including prominent scientists from the Field Museum and the nearby Morton Arboretum. The prairie was re-placed several years ago by an oil-tank storage area, yet every species of plant which Norma Pfeiffer indicated as growing with *Thismia* still occurs there. Less than one mile away, a similar habitat, known as Burnham Prairie, exists today, and *Thismia* may be hiding under the goldenrods there as well.

The fact that *Thismia americana* apparently became extinct about 1914 caused no general consternation among biologists at that time. It was assumed that certain organisms would become extinct. After all, the dinosaurs and giant horsetails and dozens of other prehistoric life forms were gone, and nearly everyone ac-cepted the fact that, given enough time, certain kinds of plants and animals would disappear naturally from the earth.

But then came the genius of man in the twentieth century. Cities exploded with phenomenal population increases, industries sprang up, mining intensified, highways were built for the new-fangled automobiles, and the natural landscape which had pro-vided homes for countless kinds of plants and animals became altered, usually with adverse effects on the living organisms. Na-tural extinctions were being overshadowed by man-caused extinc-tions.

It became apparent to some that steps would have to be taken if we wanted our children and their children to enjoy the same kinds of plants and animals we had known. Slowly, small pieces of both federal and state legislation were enacted which showed concern for our diminishing natural heritage.

Finally, the Endangered Species Protection Act of 1966 acknowl-edged a national responsibility to act on behalf of native species of wildlife which are threatened with extinction. This act provided only for mammals, birds, fish, amphibians, and reptiles to be listed in the Federal Register. As a result, on March 11, 1967, the first lists of federally endangered animals were published.

In 1973, the Endangered Species Act, widely acclaimed as a

model to be followed around the world, set up the mechanism whereby all animals and plants that warranted federal protection could be listed as endangered or threatened. For the first time, plants were placed on nearly equal terms with animals.

The 1973 act defined the terms "endangered" and "threatened," determined that the secretary of the interior implement the provisions of the act, and directed the Smithsonian Institution to review the status of nearly 25,000 kinds of native plants which grow in the United States.

According to the act of 1973, an endangered species is one which is close to extinction throughout all or a significant part of its range. A threatened species is one which is likely to become endangered in the near future.

To assist him in implementing the Endangered Species Act, the secretary of the interior delegated the Fish and Wildlife Service to accumulate and evaluate the information concerning the nation's animals and plants. Specifically, an Office of Endangered Species was created to perform these important duties.

In its response to review the status of the plants of the United States, the Smithsonian Institution, aided by information contributed by botanists from all across the nation, presented to Congress in January 1975 a list of 3,187 kinds of plants in the United States which were likely candidates for endangered or threatened status (more than a thousand of these were from Hawaii). By accepting this report in a petition appearing in the July 1, 1975, Federal Register, the Fish and Wildlife Service initiated the formal review of each of these species.

Less than one year and considerable research later, the Fish and Wildlife Service proposed on June 16, 1976, that 1,779 native plants might qualify for federal listing.

The process involved from this proposed listing of a plant in 1976 to ultimate listing as a federally endangered or threatened species required careful, detailed documentation concerning known locations for each species, historical distribution, habitat determinations, and potential threats.

Twenty-one plants had been designated for federal protection when the 1978 amendments to the Endangered Species Act required that the remaining 1,700 or so plants which had been proposed on June 16, 1976, be withdrawn since a two-year time limitation had elapsed.

Undaunted, the Fish and Wildlife Service reproposed a revised list of about 1,700 species which could ultimately make the federal list. (Another 1,100 were listed which required additional detailed study.) A total of 61 plants had been officially declared as federally endangered or threatened by late 1981, when a virtual hold was placed on further listing until the outcome of the Endangered Species Act renewal in the autumn of 1982 was known.

Sen. John H. Chafee of Rhode Island, chairman of the Senate Subcommittee on Environmental Pollution, Committee on Environment and Public Works, has pushed hard for the reauthorization of the Endangered Species Act. On March 20, 1982, he and Sens. Slade Gorton of Washington and George J. Mitchell of Maine introduced Senate Bill 2309, a bill to reauthorize the Endangered Species Act. The House Subcommittee on Fisheries and Wildlife Conservation and the Environment, Committee on Merchant Marine and Fisheries, led by Reps. John Breaux of Louisiana and Edwin Forsythe of New Jersey, introduced its version of the bill on April 20. Conservationists testified in favor of a strong endangered species act, while opponents argued to weaken the act or do away with it altogether.

Many reasons for preserving species were brought out clearly during the 1982 congressional hearings. In addition to the purely aesthetic value of living organisms, their potential as sources of food and chemicals for medicinal purposes is scarcely known. Dr. Thomas Eisner of Cornell University has estimated that only about 2 percent of the flowering plants of the world have been tested for the presence of alkaloids, a class of chemicals which often have important properties in controlling diseases.

The process of listing endangered and threatened species, slowed to a near standstill in 1981 and 1982, must be placed in high gear. William D. Blair, president of The Nature Conservancy, stated in his testimony during the reauthorization hearings that scientific research and survey work must be accelerated because of the lack of knowledge about the plants and animals in some of our states. He noted that The Nature Conservancy's natural heritage programs that exist in twenty-seven states have shown that "(a) a greater number of species of concern have been proven common, not rare; (b) new species not known to exist in a state are being discovered there; (c) species thought to be extirpated in a state are being 'rediscovered' there; and (d) targeted research has led to the discovery of species new to science." He also pointed out that conflicts are being resolved through proper use of endangered species data. Among the examples he gave was a Massachusetts Department of Transportation decision to realign a proposed highway to avoid the habitat of a rare snakeroot *(Eupatorium leucolepis* var. *novae-angliae)* before land was acquired for the project. In Arizona, a proposed airport expansion was changed to an area where there would be less impact on a sensitive environment.

On June 8, 1982, the House of Representatives passed its version of the reauthorization bill, and a day later the Senate followed suit. A joint House/Senate conference committee spent much of the summer of 1982 working out differences between the two bills before sending the final reauthorization bill to the president.

In the meantime, every state had become interested in rare plant protection, and each had some laws on the books regarding plant conservation. Many states had prepared lists of plants which were variously designated as endangered, threatened, rare, or of special concern for that particular state.

The reasons for the limited populations of the nearly 3,000 rare United States plants are nearly as diverse as the plants themselves. Some species apparently always have been rare, possibly because of a difficulty in the reproductive process, possibly because of a very precise and limited habitat requirement, or for other reasons. Some may be limited because of a natural phenomenon, such as the eruption of a volcano or a destructive hurricane. Most species, however, have diminished through the direct or indirect activities of man.

Intensive agricultural practices have destroyed millions of acres of prairies and forests, together with the plants (and animals) that once lived there. Overgrazing, particularly in the western states, has taken a great toll on plants. The search for energy sources and mineral deposits has resulted in mining operations that have wiped out some species and left others on the brink of extinction. Population growth has resulted in suburban developments and shopping centers that diminish the amount of wild habitats available for native plant and animal life, while the building of thousands of miles of roads has carved into many species' habitat.

Search for water has led to destruction of wetlands and irreparable damage to water tables. Overcollecting for commercial or personal use has caused the demise of many kinds of organisms, particularly cacti, other succulents, and carnivorous plants such as pitcher plants and fly traps. In certain fragile areas such as sand dunes, off-road vehicles have rendered the areas unsuitable for plant growth.

Since passage of the Endangered Species Act in 1973, our nation has become increasingly aware of the need to preserve our heritage, whether it be plants, animals, natural areas, archeological remains, or historical sites. It is up to each of us to act—to do our part. Let's not permit any of the 3,000 rare plants of our country to become another *Thismia.*

2

Vanishing Wildflowers
of the Northeastern States

The sixteen rare plants discussed in this chapter include three which are currently on the federal list of endangered and threatened species, one other which has been proposed for federal listing, and twelve which are under review by federal or state agencies. In addition, the northern monkshood, which occurs in New York, Ohio, Wisconsin, and Iowa, is a federally threatened species that is discussed in Chapter 4 on the north-central states.

The states that are included in this chapter are Maine, New Hampshire, Vermont, Massachusetts, Rhode Island, Connecticut, New York, New Jersey, Delaware, Pennsylvania, Maryland, and West Virginia.

Furbish's Lousewort or *Pedicularis furbishiae.*
Snapdragon family. Perennial; stems to 18 inches. Leaves alternate, deeply divided. Flowers greenish yellow, in crowded terminal spikes; petals 2-lipped, snapdragonlike, ¾ inch long. Season: July, August.

Little could Kate Furbish realize that the lovely lousewort which she discovered in 1880 on the banks of Maine's St. John River would become the most controversial plant in the United States a century later.

Pedicularis furbishiae is one of more than 300 louseworts found in North America, Europe, and Asia. They derive their name from the old belief that cattle would become heavily infested with lice if they ate plants of this genus.

During the sixty-three years following its original discovery,

Furbish's lousewort was found a number of times along the St. John River in northern Maine and adjacent New Brunswick, Canada. In 1943 the last collection was made of this species for more than three decades, and it was generally regarded as extinct.

The decline of this species was due to several factors. Much of the area where it grew was cleared to make room for potatoes, one of the most important crops of Aroostook County, Maine. Pulp mills were built along both sides of the St. John River, clearing more land where Furbish's lousewort lived. Homes built along the river must have destroyed some of the plants. Even nature played a role in the deterioration of the lousewort's habitat by periodically scouring and cutting away the banks of the river.

In the meantime, the United States Army Corps of Engineers was laying plans for a huge hydroelectric project on the St. John River which would provide for a less expensive source of energy for residents of the area. The Dickey-Lincoln hydroelectric project called for an earthen-filled dam 3 miles long and 900 feet high, backing up the St. John River for 50 miles and flooding 86,000 acres of valley forest. Plans for the project were sailing along until the summer of 1976, when Dr. Charles S. Richards, a botany professor from the University of Maine, made the startling rediscovery of Furbish's lousewort along the St. John River.

The fervor among conservationists and the press about the rediscovery of this "extinct" species and its imminent destruction by an impending dam prompted the Corps of Engineers to fund research on the plant. Dr. Richards was selected to conduct a search for more locations for this species.

Dr. Richards succeeded in finding 28 colonies of Furbish's lousewort, with a total of 5,424 flowering stems. More than half of these were in the area that would be inundated when the hydroelectric project was completed.

Pedicularis furbishiae occupies a unique niche along the St. John River. The banks of the river are terraced. At the top of the steep bluffs are dense forests of white spruce, punctuated occasionally by a picturesque paper birch. The bluff falls away to a ten-foot-wide terrace of sandy loam and gravel. It is in this narrow zone that Furbish's lousewort grows among a rich assemblage of fireweed and loosestrife overtopped by a dense thicket of downy alder.

Researchers at the University of New Hampshire have been working on the possibility of propagating Furbish's lousewort by cloning, a technique that is highly successful with orchids.

Meanwhile, the United States Fish and Wildlife Service and the State of Maine are cooperating in an awareness program designed to alert landowners in the St. John River valley as to the significance of the plant and the need for protecting it.

What kind of plant is Furbish's lousewort? It is a member of the snapdragon family and is a perennial with bright yellow roots. Above ground an unbranched purple stem rises to about eighteen inches tall, with many deeply divided leaves scattered along it.

At the tip of each stem is a spike of six or more flowers, each flower surrounded by and at first surpassed by leafy bracts. As the snapdragonlike flowers expand to their maximum length of three-fourths of an inch, they protrude beyond the bracts. The flowers bloom during July and August.

Because of the extremely limited range and the various threats to its existence in the area, Furbish's lousewort was listed as a federally endangered species on April 26, 1976.

Attention to this species was intensified during 1977 when CBS newsman Charles Osgood described its plight in his inimitable style of verse:*

> *Kate Furbish was a woman who a century ago*
> *Discovered something growing, and she classified it so*
> *That botanists thereafter, in their reference volumes state,*
> *That the plant's a Furbish lousewort. See, they named it after Kate.*
> *There were other kinds of louseworts, but the Furbish one was rare.*
> *It was very near extinction, when they found out it was there.*
> *And as the years went by, it seemed with ravages of weather,*
> *The poor old Furbish louseworts simply vanished altogether.*
> *But then in 1976, our bicentennial year,*
> *Furbish lousewort fanciers had some good news they could cheer.*
> *For along the St. John's River, guess what somebody found?*
> *Two hundred fifty Furbish louseworts growing in the ground.*

* From *Nothing Could Be Finer Than a Crisis that Is Minor in the Morning* by Charles Osgood. Copyright © 1979 by CBS, Inc. Reprinted by permission of Holt, Rinehart and Winston, Publishers.

Now, the place where they were growing, by the St. John's River
banks,
Is not a place where you or I would want to live, no thanks.
For in that very area, there was a mighty plan,
An engineering project for the benefit of man.
The Dickey-Lincoln Dam it's called, hydroelectric power.
Energy, in other words, the issue of the hour.
Make way, make way for progress now, man's ever constant urge.
And where those Furbish louseworts were, the dam would just
submerge.
The plants can't be transplanted; they simply wouldn't grow.
Conditions for the Furbish louseworts have to be just so.
And for reasons far too deep for me to know or to explain,
The only place they can survive is in that part of Maine.
So, obviously it was clear, that something had to give,
And giant dams do not make way so that a plant can live.
But hold the phone, for yes they do. Indeed they must, in fact.
There is a law, the Federal Endangered Species Act,
And any project such as this, though mighty and exalted,
If it wipes out threatened animals or plants, it must be halted.
And since the Furbish lousewort is endangered as can be,
They had to call the dam off; couldn't build it, don't you see.
For to flood that lousewort haven, where the Furbishes were at,
Would be to take away their only extant habitat.
And the only way to save the day, to end this awful stall,
Would be to find some other louseworts, anywhere at all.
And sure enough, as luck would have it, strange though it may seem,
They found some other Furbish louseworts growing just downstream.
Four tiny little colonies, one with just a single plant.
So now they'll flood that major zone, no one can say they can't.
And construction is proceeding, and the dynamite goes bam.
And most folks just don't seem to give a Dickey-Lincoln Dam.
The newfound stands of Furbish louseworts aren't much, but then
They were thought to be extinct before, and well may be again.
Because the Furbish lousewort has a funny-sounding name,
It was ripe for making ridicule, and that's a sort of shame.
For there is a disappearing world, and man has played his role
In taking little parts away from what was once the whole.
We can get along without them; we may not feel their lack.

But extinction means that something's gone, and never coming back.
So, here's to you, little lousewort, and here's to your rebirth.
And may you somehow multiply, refurbishing the earth.

Dwarf Cinquefoil or *Potentilla robbinsiana*.

Rose family. Tufted perennial; stems to 2 feet, densely hairy. Leaves alternate, the lowest divided into 3 coarsely toothed leaflets ⅓ inch long. Flower solitary, yellow, ¼ inch across; petals 5. Season: June, July.

My first visit to New Hampshire's Mount Washington as part of a combined working vacation for my family and me was a memorable one. I was engaged in a research project in the White Mountain National Forest during the summer of 1973. On one occasion I wished to examine the alpine flora above timberline in the Presidential Range. My wife, three children, and I decided that a trip up the cog steam railway would be a pleasant respite from the day-by-day hiking and plant hunting we were used to.

As the train climbed beyond Jacob's Ladder, the steepest grade on the railroad, we were met with waves of fog. Although it was summer, the temperature when we disembarked from the train was barely fifty, and by the time we breathlessly trudged to the Summit House, we were unable to make out the green and red engine that had pushed us to the top.

After refreshments, we cautiously made our way along the gravelly, nearly barren trail that led away from Mount Washington's summit in the direction of Mount Monroe. Because of the severe climate of this mountain range, the growing season for plants is limited to less than four months. The snow which blankets the ground for such a long winter is beneficial to the plants, sheltering them during the winter and providing moisture when it melts to usher in the growing season. Many of the alpine plants in the Presidential Range are extremely northern species which grow only as far south as Mount Washington.

The plant we particularly wanted to see that day was the dwarf cinquefoil, *Potentilla robbinsiana*, a yellow-flowered member of the rose family. This small plant had originally been found in 1829 on Mount Monroe, and apparently it was the same colony that was still in existence when we observed it in 1973.

Several plants were growing in dense tufts. The many, crowded

Furbish's Lousewort
Pedicularis furbishiae

Silverling
Paronychia argyrocoma
var. *albimontana*

Frostweed
Helianthemum dumosum

Dwarf Cinquefoil
Potentilla robbinsiana

leaves, each hairy and divided into three leaflets, are only about two inches long. Rising occasionally from among the leaves is a two-inch stalk bearing a solitary, terminal yellow flower in late June or early July.

The dwarf cinquefoil has been reduced in numbers during the fifteen decades it has been known because of its nearness to the well-traveled path which bisected the original colony. The old Crawford Path Bridle Trail of the 1860s in modern times has been converted into a part of the Appalachian Trail. Although some plants have actually been trampled to death, mere disturbance near the plants loosens the soil, which then is easily blown or washed away.

James Robbins, a pioneer New England plant enthusiast as well as a physician, was the first person to discover this species on Mount Monroe. Sometime late in the nineteenth century, the dwarf cinquefoil was found on Mount Mansfield, Vermont. During 1897, 1915, and 1963, this species was found in two places in New Hampshire's Franconia Range, centering on Mount Lincoln. Recent searches, however, have failed to reconfirm plants at either Mount Mansfield or Mount Lincoln. In 1970, however, a small colony was located on private property in adjacent Vermont by Donald White, who, at the age of seventy-seven, had been a member of the New England Botanical Club for fifty-nine years.

The dwarf cinquefoil was listed as a federally endangered species on September 17, 1980. With this designation, the plant should receive the protection which may prevent its extinction. The Appalachian Mountain Club, the United States Forest Service, the Fish and Wildlife Service, and the University of New Hampshire have been planning a proposed route change for a portion of the Appalachian Trail which would direct hikers away from the cinquefoil.

Small Whorled Pogonia or *Isotria medeoloides.*
Orchid family. Perennial; stems to 8 inches. Leaves 5 or 6 in a single ring on the stem, elliptic, to 3 inches. Flowers few, greenish yellow, 1 inch long. Season: May, June. (See color plate 1.)

An article in a prominent conservation magazine a few years ago referred to the small whorled pogonia orchid (*Isotria medeoloides*) as the rarest orchid in America. Whether this tiny plant of the eastern United States deserves that unique distinction is open to question, but there is no doubt that this orchid is among the rarest in the nation and deserves total protection.

Although Frederick Pursh first named this orchid from Virginia in 1814, it was generally not found again until the turn of the century. In the nearly 170 years this species has been known, it has been found barely fifty times in sixteen states. Many of these stations are no longer in existence, and at most of the others, only a very few plants are growing.

The orchid genus *Isotria* is comprised of only two species, both confined to the eastern United States. The larger of the two, *Isotria verticillata*, is itself a fairly rare plant, but is in little danger of becoming extinct.

Both Isotrias are similar in that they bear a single ring of leaves at the top of the stem. From this group of leaves arises one or two flowers.

The flower is made up of three narrow green sepals about one inch long, two light green to greenish yellow petals that are shorter than the sepals, and a third petal, the lip. The lip is greenish white, about one-half inch long and one-third as wide, and three-lobed.

Michael Homoya of the Indiana Natural Heritage Program made many careful observations on *Isotria medeoloides* while studying it in Illinois for part of his master's research at Southern Illinois University. Homoya found that once the plant broke dormancy (at least where it could be observed breaking through the soil), the stem elongated very rapidly, the leaves expanded, and the flower or flowers opened, all within a period of eight days. Four days later, the flowers wilted. If fertilization occurred, the ovary began to expand into the capsule, and the old flower stalks lengthened slightly.

The flowers bloom as early as the first week in May in the southern part of this orchid's range, while in the northern states it may not bloom until early June.

In most of the areas where the small whorled pogonia has been found, it grows in a rather dry, oak-dominated woods in thick leaf mold. However, in Massachusetts it was reported as growing under hemlock, while in Michigan it occurs in low, sandy woods. At the Illinois locality, the leaf mold on the ground is not thick, and the habitat is a steep slope above a vertical sandstone cliff.

The largest colony of this orchid known in the world today is in Maine, where 190 plants were counted in 1981. There are more plants here than at all the other locations put together. The area has been registered as a critical area in Maine.

There are a number of isolated sites in Pennsylvania for this orchid, and perhaps at least four known localities each in New Jersey and New Hampshire.

In New York, this orchid has been found at five localities in four counties, but since 1930, only a single station has been known to exist in the state, and it consists of one plant which appears about every three years.

Connecticut has had the most historical sightings for *Isotria medeoloides,* and it is believed to be extant in that state today, but it is very rare. Currently there are eight plants known to exist in Rhode Island.

In the southeast, the small whorled pogonia is known from a few localities in North Carolina, South Carolina, and Virginia.

In Randolph County, Illinois, where this orchid was found in 1973, there are two to five plants, depending on the year, since each plant has not come up annually during the time this population has been studied.

The Michigan plants were collected in 1968 in a low woods that had formerly been an orchard.

Isotria medeoloides is apparently extirpated from Missouri, where it was found once in 1897; from Massachusetts, where it was collected only in 1899; from Vermont, where it has not been seen since 1902; and from Maryland, where it held on in Montgomery County until 1930.

Homoya has suggested four reasons that the small whorled pogonia orchid is rare. It may reproduce inadequately; it may be unable to compete successfully with other plants for light, moisture, and nutrients; its habitat may have been destroyed through man's activities; or it is a senescent species, that is, a very ancient plant that has not become a successful member of our modern flora.

On September 10, 1982, the United States Fish and Wildlife Service added this to the list of endangered species.

Silverling or *Paronychia argyrocoma* var. *albimontana*.
Carnation family. Tufted perennial; stems to 12 inches. Leaves opposite, very narrow, to 1 inch. Flowers in dense clusters to 1 inch across; bracts silvery; sepals 5; petals none. Season: June to September.

The dictionary on the desk before me has the following entry:

sil·ver·ling (sil′ver ling), *n.* 1. an old standard of value in silver. 2. a piece of silver money: *There were a thousand vines at a thousand silverlings* (Isaiah 7:23).

An addition to this entry is in order. It should state:

3. A flowering plant of New England (*Paronychia argyrocoma* var. *albimontana*), belonging to the carnation family, on the verge of extinction and needing protection.

The word silverling has a pretty ring to it and is appropriately applied to a handsome little tufted plant confined to a few mountains in Maine and New Hampshire and to a rocky island in Massachusetts.

There are actually two distinct variations of the silverling. One of them, referred to as the typical variety, is found on mountains in West Virginia, Virginia, North Carolina, Tennessee, and Georgia. It is not endangered. The other variety, known as var. *albimontana*, is confined to perhaps a dozen sites in northern New England. It is threatened with extinction and is the subject of this account. The more common variety of the southeastern states has larger flowers and hairier leaves than the New England variety.

The silverling is a tufted perennial that grows in colonies numbering from one plant to more than one hundred plants. Although each plant grows from a single, slender taproot, the stem usually branches at ground level and lies on the rock surface, often sending up erect stems to a height of four to twelve inches. Several flowers are crowded into dense clusters which top the upright branches. Each cluster of flowers may be an inch across and is conspicuously silvery-shiny, accounting for the common name. The flowers themselves are much reduced and consist only of sepals, stamens, and a pistil; there are no petals.

The New England silverling occurs primarily on bare granitic ledges below 4,000 feet in the mountains of New Hampshire and Maine. Edward Tuckerman, who first began exploring the Presidential Range of New Hampshire in 1837 at age twenty, was the first collector of the silverling, finding it on Mount Clinton early in the 1840s. Since that time it has been found at eighteen additional sites in the White Mountains, but fewer than a dozen locations are known to exist in these mountains today.

In Maine, only three of the seven historical localities of the silverling exist. The decline of this plant has been due to human disturbance by intensive hiking in the fragile mountain habitat and by overcollecting.

A surprising discovery of the silverling was made in 1945, when 196 colonies were found on a narrow ledge on an island in the Merrimack River in Massachusetts. By 1980, this population had been reduced to 104 colonies.

On October 27, 1980, the Fish and Wildlife Service gave notice that it was proposing the silverling as a threatened species. By mid-1982, this proposal had not been acted upon.

Frostweed or *Helianthemum dumosum.*
Rockrose family. Tufted perennial; stems to 15 inches. Leaves alternate, yellow-green, elliptic, to ⅔ inch. Flowers of two kinds, the early ones with 5 yellow petals to ½ inch long, the later ones without petals. Season: May and June for the early flowers, July to September for the later ones.

Frostweeds derive their common name from a strange but beautiful phenomenon that occurs early in the morning of the season's first frost. At this time of year, the tops of the frostweeds are dead and gone, but the still-active root system causes the remaining plant sap to ooze from the broken stem bases. With air temperatures in the twenties, the sap freezes along its inner surface, pushing the ice outward. Because of the irregular cracking of the stem bases, the ice which forms is a ribbon of pure, clear ice alternating with areas of aerated, white ice. Where several of these ribbons arise from the same area, the effect of a flower made from ice is given. Only a few kinds of plants are known to do this; it is most often observed in *Helianthemum.*

There are more than 125 different kinds of *Helianthemum* in the world, although some of them are usually referred to as sun roses or rock roses, rather than frostweeds. Several occur in the United States, including four in New England. One of the rarest of these is *Helianthemum dumosum*. All of the New England ones live in dry, sandy soil, sometimes in woods or sometimes in the open. New Yorkers often refer to the habitat as "dry downs." When Merritt Lyndon Fernald of Harvard University found *Helianthemum dumosum* near Harwich, Massachusetts, in 1918, he described it as "forming depressed mats in open sandy spots in woods."

Eugene P. Bicknell, who discovered and named several plants from the eastern United States, was the first person to realize that *Helianthemum dumosum* was a new species. He had been observing plants in sandy banks on Nantucket Island, Massachusetts, when he noticed that a few frostweeds were blooming about a week before the others. When the later ones flowered, Bicknell compared specimens of the two and found that there were structural differences as well. The later-flowering plant was the rather common Canadian frostweed *(Helianthemum canadense)*. The earlier-flowering plants were an undescribed species which Bicknell officially named in 1913.

Several Massachusetts localities for this species have been found since Bicknell's collections, and the plant still exists on Cape Cod and Nantucket Island. In New York, this rare frostweed has been declining rapidly because of the destruction of the dry downs habitat due to development. New York botanists were able to locate only three stations for this species at the eastern end of Long Island during a recent intensive survey.

Helianthemum dumosum has been known from Rhode Island since its discovery on Block Island in 1913. Only recently, Richard L. Champlin, an ardent naturalist from Jamestown, Rhode Island, has found it in a different location in his state.

In Connecticut, this rare species is limited to sandy areas in the Southeast Hills and South-central Lowlands regions of the state.

During the first part of the growing season, the stems of this perennial usually form loosely matted mounds, but as the summer progresses some of the slender, somewhat woody stems become upright. An unusual feature about this species (as well as some of the other species of *Helianthemum*) is that it forms two different kinds of flowers on the same plant. The first flowers to bloom during late May and June at the tip of each branch are yellow and rather showy. From July to September, much reduced, inconspicuous flowers that have no petals are borne at the ends of the branches.

Because of the limited range of *Helianthemum dumosum*, and because of the rapid destruction of its sandy habitats, this species is in jeopardy and should be protected wherever it occurs. It has no federal status as yet.

Boott's White Lettuce or *Prenanthes boottii*.

Aster family. Perennial; stems to 12 inches, soft-hairy above. Lowest leaves heart-shaped at base, sometimes toothed. Flowers 9–15 in heads, with small, white, strap-shaped rays. Season: June, July.

Boott's white lettuce, *Prenanthes boottii*, is another plant that grows near or on mountaintops in New England. Although several people had climbed Mount Washington since Darby Fields's conquest of its summit in 1642, no one had collected botanical specimens for scientific purposes until Dr. James Bigelow and Dr. Francis Boott ascended via the Cutler River route in 1816. On that expedition, several plants were collected, including one which was later to become formally described by the great American botanist, Asa Gray, as *Prenanthes boottii*, or Boott's white lettuce.

For more than a century and a half, naturalists have observed this interesting member of the aster family at several locations in alpine regions of the Presidential Range of New Hampshire. In fact, the plant had been witnessed so many times that Arthur Stanley Pease in his *Flora of Northern New Hampshire* in 1954 refers to it as common. "Locally abundant" might be a better phrase to indicate its occurrence in the Presidential Range, meaning that it isn't found in too many places, but where it does occur, it is common. But the Presidential Range is the only area in New England where Boott's white lettuce is locally abundant, for it is known from only two mountaintops in Vermont, two locations in New York, and probably no more than two in Maine (although it used to occur in at least a third Maine locality).

I have often wondered how a plant could survive so near to the top of Mount Washington because of all the snow. Mount Washington's summit annually averages about 220 inches of winter snow. During the winter of 1968–69, 566.4 inches of snow were recorded! What explanation might there be? The fact is that every snowfall on Mount Washington is usually followed by a period of heavy winds, often with a velocity of seventy to eighty or more miles per hour. The winds blow the snow off the mountaintop and down into ravines, where it accumulates. Wesley Tiffany, a student at the University of Massachusetts at Boston in 1967, found that six days following a recorded 22 inches of snow in mid-February at the Mount Washington summit weather station, the area was nearly devoid of snow because of the action of the wind.

Boott's white lettuce occurs in rocky, tundra habitat, usually just above a zone of spruce-fir forest. The hazards to it are similar to those of the dwarf cinquefoil, namely disturbance by hikers and overcollecting by naturalists. At one of its two locations in Vermont, the mountaintop receives about 10,000 visitors each year.

The plant stands no more than a foot tall. The upper half of each plant has an elongated series of flowering heads. Each head, which is composed of nine to fifteen flowers, droops. By mid-July each head has become conspicuous because many soft, tawny bristles which are attached to each seed have enlarged and become rather showy. Eventually, the wind will blow each head apart, and the seeds will be carried by means of these bristles, which serve as a parachute.

Boott's white lettuce is not yet on the federal list of endangered or threatened species, but it is uncommon enough that it should be protected wherever it is found.

Spreading Globe Flower or *Trollius laxus.*

Buttercup family. Perennial; stems to 2 feet. Leaves deeply 5- to 7-lobed, the lowest on long stalks. Flower solitary, greenish yellow, to 1½ inches across; sepals 5–15, petallike; petals none; stamens numerous, some of them petallike. Season: April, May. (See color plate 2.)

The spreading globe flower, *Trollius laxus,* is a lovely member of the buttercup family. It belongs to a genus of perennial herbs that is found in the north temperate regions of both the New and the Old World. The European globe flower is often grown as a garden ornamental in the United States.

Unlike many of the rare species in this country which have very restricted ranges, the spreading globe flower has been found from Connecticut to Ohio, with intermediate localities in New York, New Jersey, and Pennsylvania. Merritt Lyndon Fernald, the indefatigable botanist from Harvard University who died in 1950, stated that there are old records of this species from western Maine and western New Hampshire. (Continued reports of this plant from Michigan are apparently in error, according to Dr. Edward Voss, the foremost authority on the Michigan flora.)

Spreading globe flower grows in very moist situations. Collectors have noted it from rich meadows, swamps, wet woods, swales, and open calcareous swamps. At nearly every location where the *Trollius* had been known to occur in the northeastern United States, it has either completely disappeared or has declined in number. New York botanists who searched their state for this species in 1979 have reported that it is apparently gone from over half the sites visited, and that it no longer occurs in southern New York. At one time the spreading globe flower had been found in seventeen New York counties in as many as forty-eight different locations. In Connecticut, *Trollius laxus* occurs in a few swamps in the northern Marble Valley region. New Jerseyites can still find this plant in three counties, and Pennsylvanians may be fortunate enough to have it in seven counties. In Ohio, the spreading globe flower was found as early as 1835 near Canton.

This delicate plant seems to be extremely sensitive to disturbances. Areas where it once occurred and which later were used as pastures no longer support this plant. A few of its former sites have been drained to permit housing developments. Because of its inherent need for adequate moisture, this species quickly succumbs when its water supply is diminished. There is also no doubt that some native plants have been dug up and transplanted to gardens because of the attractive flowers.

This species is an erect perennial growing to a height of one to two feet. Usually only one greenish-yellow flower terminates each stem; it is about one and one-half inches across. When the flower begins to open, the sepals are more or less concave and ascending, causing each flower to be shaped like a small globe. The flowers bloom from late April and into May. The fruits mature shortly thereafter.

The famous botanist Gotthilf Henry Ernest Muhlenberg was apparently the first person to discover this species. In 1791, he named this species on the basis of plánts he found growing in the Lancaster, Pennsylvania, area.

The spreading globe flower is under consideration for possible listing as an endangered or threatened species.

Tidal Shore Beggar's-Tick or *Bidens bidentoides.*

Aster family. Annual; stems branching, to 3 feet. Leaves alternate, lance-shaped. Flowers crowded together into several heads up to 1 inch across, without any petallike rays. Season: August to October.

Bidens is known as beggar's-tick, because the dark seeds which can become embedded in clothing sometimes resemble ticks in shape and color. These seeds are able to become attached to clothing by means of a pair of sharp projections equipped with usually downward-pointing barbs.

There are many species of *Bidens* in the United States, and most of them are wide-ranging and common. Several of them have showy flower heads with bright yellow rays. *Bidens bidentoides* does not follow this pattern. It is a rare species restricted to Pennsylvania, Delaware, Maryland, New Jersey, and New York, although it has not reached threatened status in New York as yet. It also has inconspicuous flower heads because there are no yellow rays present.

This *Bidens* grows in fresh to brackish tidal shores and on mud flats. Since a large part of the tidal flats where it occurs is being altered for erosion control, this species is becoming increasingly endangered.

It is a completely smooth annual with a branched stem reaching a height of about three feet. Each head, which is less than an inch across, is surpassed by a few green, narrow bracts. None of the individual flowers within the head has rays. The "ticks," which are the fruits, are narrow and taper to the base.

In New York, this species grows occasionally on the tidal shores of the Hudson River. Bayard Long collected this beggar's-tick in muddy shores of the Maurice River in New Jersey in 1909. He found it again in 1935 on a tidal shore in the vicinity of Manantico.

In Pennsylvania and Delaware, *Bidens bidentoides* grows in fresh tidal mud along the Delaware River, although it has not been seen in Delaware since 1946.

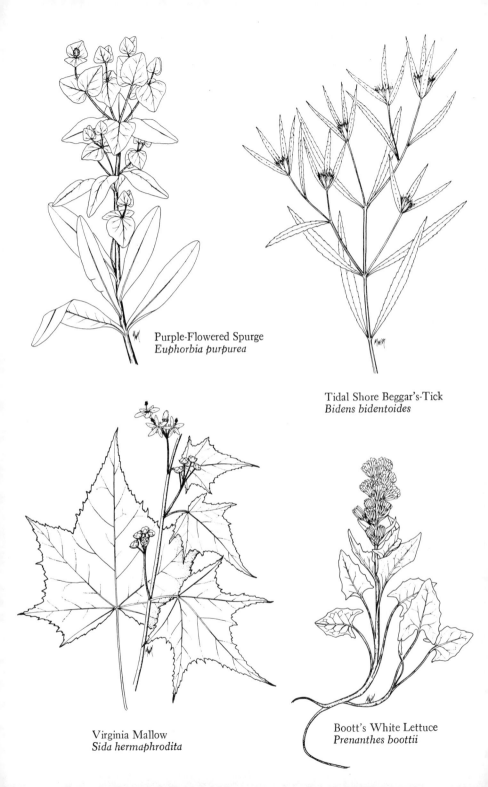

Purple-Flowered Spurge
Euphorbia purpurea

Tidal Shore Beggar's-Tick
Bidens bidentoides

Virginia Mallow
Sida hermaphrodita

Boott's White Lettuce
Prenanthes boottii

Finally, in Maryland, this rare beggar's-tick grows on mud flats on the shores of upper Chesapeake Bay and a few of its tributaries.

Purple-Flowered Spurge or *Euphorbia purpurea.*
Spurge family. Perennial; stems to 3 feet. Leaves alternate, oblong, without teeth, softly hairy, to 3 inches. Flowers small, cup-shaped, purple, borne at the ends of slender rays. Season: July to September.

The spurge that Rafinesque saw on the Allegheny Mountain glade in Pennsylvania in 1838 was different. He had never seen anything like it in the nearly twenty-five years that he had been exploring the eastern United States. To this new species he gave the name *Euphorbia purpurea*, the purple-flowered spurge.

Constantin Samuel Rafinesque-Schmaltz, a friend of John James Audubon, had come to this country in 1815 from Sicily, where he had been employed as a merchant. But his unquenchable thirst for plant exploration and collecting, which had been developing since he was eleven years old, soon won out over a career as a merchant. From 1815 until the time of his death, he explored much of the Ohio Valley on foot, making many discoveries of new plant species. For a while he served as professor of natural history at Transylvania University in Lexington, Kentucky.

Spurges belong to the genus *Euphorbia*, and there are hundreds of different kinds in the world. Some of them are leafless and succulent, often being mistaken for cacti. Others, like the famous crown-of-thorns, are beset with vicious spines. A few are well known because of their showy nature, such as the poinsettia, snow-on-the-mountain, and cypress spurge. Many spurges are obscure members of the flora, some occurring in native habitats, others in disturbed situations.

Despite the great variation found among the different kinds of spurges, all have some things in common. They possess latex, or milky sap, and their flowers lack petals. (The showy parts of a poinsettia "flower" are actually colored bracts, not petals.)

The spurge Rafinesque found in 1838 was a perennial with a short, thick, underground rhizome. Above ground, the stems grow to a height of three feet or more. Near the top of each stem, several leaves form a ring around the stem. From among the bases of these leaves arise several very tiny, cup-shaped, deep purple flowers.

The purple-flowered spurge may occur on mountain glades, where Rafinesque first found his specimen, or it may grow in rich or swampy woods and thickets.

Several nineteenth-century botanists observed it in Pennsylvania after Rafinesque's discovery. The species occurs today in Chester, Cumberland, and Lancaster counties.

In 1891, the purple-flowered spurge was collected in New Castle County, Delaware, but this plant apparently no longer occurs in Delaware. Maryland localities in Cecil and Carroll counties have not been refound since 1928 and 1940, respectively.

Euphorbia purpurea is also known today from a few areas in West Virginia, Virginia, and North Carolina, although North Carolina botanists consider it endangered in their state.

The general rarity of this species throughout its range makes it a candidate for future federal listing.

Virginia Mallow or *Sida hermaphrodita.*

Mallow family. Perennial; stems smooth, to 13 feet. Leaves alternate, deeply lobed, to 6 inches wide. Flowers white, 1 inch across, in clusters at the ends of branches; petals 5. Season: July to October.

Sida hermaphrodita, the Virginia mallow, was described by the great Swedish scholar Carolus Linnaeus in 1753, based on specimens sent to him from the eastern United States. Linnaeus received these plants at about the same time that he obtained specimens of a similar-looking mallow from Virginia. Linnaeus thought that the two different plants were closely enough related to be placed in the same genus, which he named *Napaea*. One of the species had both stamens and pistils in the same flower. Linnaeus appropriately called it *Napaea hermaphrodita*. The other one had the stamens and pistils in separate flowers on separate plants, a condition referred to by botanists as dioecious. This species Linnaeus called *Napaea dioica*, and is the glade mallow known today from the eastern and central United States. It was later determined that *Napaea hermaphrodita* wasn't really a *Napaea* at all, but was more closely allied to another genus in the mallow family called *Sida*. Consequently, H. H. Rusby of the New York Botanical Garden changed the name of this species in 1894 to *Sida hermaphrodita*, which it is known by today.

While some rare plants may simply be overlooked because of their small stature, the same cannot be said for the Virginia mallow. This coarse, nearly smooth perennial may grow to a height of thirteen feet. Clusters of white flowers arise from the ends of the branches during the summer.

Because the Virginia mallow is attractive enough to be grown in gardens, it is sometimes difficult to tell whether a collection has been made in the wild or from a garden, since some of the earlier collectors never indicated this important fact. I have seen specimens from Illinois, Michigan, and Massachusetts which probably came from planted individuals.

The native range of the Virginia mallow is thought to center on three disjunct areas. One of these is in the Potomac and Susquehanna rivers' watersheds in Pennsylvania, Maryland, and the District of Columbia; one is in extreme southern Ohio and adjacent southwestern West Virginia; and a third is in the far northeast corner of Tennessee.

In 1972, L. K. Thomas, a research biologist with the National Park Service, conducted an extensive study of the Pennsylvania–Maryland–District of Columbia populations. He found that during the last one hundred years, approximately two-thirds of the populations in that area had been extirpated. At those sites where the plants were still observed, they were growing in loose, sandy or rocky soil along riverbanks or in ravines. In at least two areas, the habitat is a floodplain woods which is probably inundated annually. The Virginia mallow appears to be able to tolerate somewhat disturbed conditions.

In West Virginia, *Sida hermaphrodita* occurs only in the Kanawha drainage, where it is found along riverbanks and in woodland openings. The Ohio colonies are found on sandy banks of the Ohio River in at least four counties.

The Tennessee collections are puzzling. They were all collected before the turn of the century by Tennessee's most famous resident collector at that time, Dr. Augustin Gattinger. Dr. Gattinger found this species in the Cumberland Mountains along the Tennessee-Kentucky state line. It is interesting to note that when Gattinger later published a list of Tennessee plants, he failed to include his own specimens from the Cumberland Mountains.

Although this species seems to be getting scarce, it is not yet listed as an endangered or threatened species in West Virginia,

Ohio, or Tennessee. Nonetheless, its general rarity calls for protective measures.

Sensitive Joint-Vetch or *Aeschynomene virginica.*
Pea family. Annual; stems much branched, to 8 feet. Leaves alternate, compound, divided into 70 or more narrowly oblong leaflets. Flowers yellow, sweet pea-shaped, ½ inch long, in elongated clusters from axils of leaves. Pod 1 inch long, hairy. Season: August to October.

The sensitive joint-vetch, *Aeschynomene virginica*, has had an interesting history of collections. The plant was first discovered by Virginia's eighteenth-century botanist, John Clayton, who found it along the Rappahannock River. This member of the legume family was not seen again in Virginia for 200 years, a fact which disturbed Merritt Lyndon Fernald of Harvard University. In 1938, Fernald set out to look for this species and others along the tidal shores of eastern Virginia. He described in the botanical journal *Rhodora* his thrill in relocating this long-lost Virginia plant on the tidal shores of the James River below Scotland. Much to his surprise, Fernald found the plant several times during the summer of 1938. The following year, he saw it so many times that he no longer became excited when he found it towering over colonies of handsome-flowering beggar's-ticks.

Aeschynomene virginica is holding its own today in the tidal marshes of Virginia and North Carolina. It is not considered endangered or threatened in either state. The same cannot be said for this species elsewhere, however.

After Clayton's original discovery of this species in Virginia, the sensitive joint-vetch was found on fresh to brackish tidal shores in Delaware, Maryland, New Jersey, and Pennsylvania. It was known at least as early as 1896 from the Wilmington, Delaware, area, but the plant apparently is gone from that state today. The same is probably true in Pennsylvania, where sites on sandy or muddy riverbanks in Delaware and Philadelphia counties could not be located during a recent survey.

In Maryland, the sensitive joint-vetch is found in Somerset County, although earlier collectors had obtained it in Anne Arundel, Prince Georges, and Wicomico counties.

Wayne Ferran, Jr., of the Academy of Natural Sciences of Philadelphia, has documented the occurrence of the sensitive joint-vetch in recent years along the Maurice River system in New Jersey. It is not likely that this species would be overlooked. Although an annual, its partially hairy, bushy-branched stems may grow as tall as eight feet. Beginning in July and continuing into October, a few yellow, sweet-pea-shaped flowers with red veins are formed in elongated clusters.

Because of the moderate occurrences of the sensitive joint-vetch in Virginia and North Carolina, this plant is not on the federal list of endangered and threatened species. Its restricted habitat makes continuous monitoring of this species imperative, however.

Depauperate Aster or *Aster depauperatus*.

Aster family. Perennial; stems much branched, smooth, to 1 foot. Leaves alternate, variable in shape and size from spatula-shaped to threadlike to very short and narrow. Flowers crowded together in small heads, with very short white rays surrounding a central cluster of small, tubular flowers. Season: August to October.

Serpentine barrens is a name applied to a very restricted habitat found in a small area in parts of Chester, Delaware, and Lancaster counties, Pennsylvania, and Cecil County, Maryland. The rocks which outcrop in the serpentine barrens are composed of the mineral serpentinite, a substance found in various places throughout the world. It takes its name from the fact that on certain serpentinite outcrops in northern Italy there lives a green and brown snake which blends in perfectly with the rocks on which it lives.

Wherever serpentinite occurs, the vegetation is usually sparse because of toxic amounts of magnesium, nickel, and chromium. Usually lacking in serpentinite are adequate amounts of calcium, potassium, and phosphorous, necessary for good plant growth.

Although the vegetation is sparse on the serpentine barrens, the species that occur are usually dissimilar from those in surrounding habitats.

The plant of the serpentine barrens with the most restricted range is the depauperate aster (*Aster depauperatus*). Belonging to that great and massive aster family, this species is known to exist

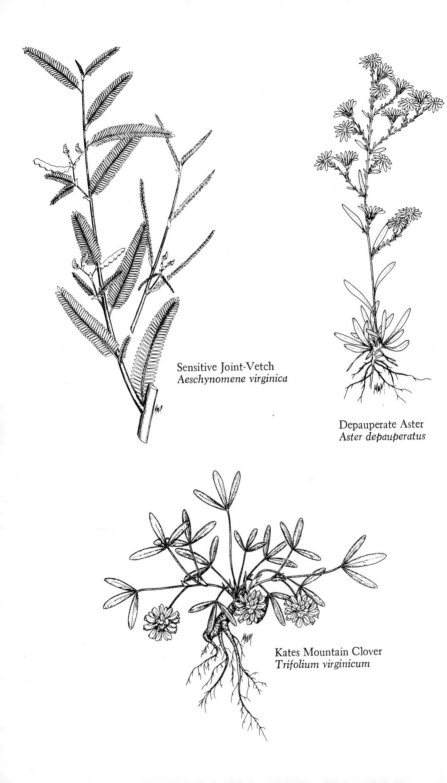

Sensitive Joint-Vetch
Aeschynomene virginica

Depauperate Aster
Aster depauperatus

Kates Mountain Clover
Trifolium virginicum

in natural habitats only from three counties in Pennsylvania and one in Maryland. Reports of it from Delaware and West Virginia as early as 1950 are apparently erroneous.

Depauperatus is a well-chosen epithet for this species, because its characteristics are diminutive when compared to those of most other asters. This aster stands only about one foot tall. The flowering head is typically aster, with a series of green bracts surrounding a number of short, white rays, with a few tubular disk flowers in the center of the head. As the flower head matures, these small rays usually turn pinkish. Blooming time for this species spans August, September, and October.

Although the depauperate aster has no official status as an endangered or threatened species, botanists in both Pennsylvania and Maryland consider it rare or local. With its distribution severely restricted to only four counties in the world, it should be carefully protected.

Kates Mountain Clover or *Trifolium virginicum.*

Pea family. Perennial; stems sprawling, slender, hairy. Leaves alternate, divided into 3 very narrow leaflets to 3 inches long. Flowers white, sweet-pea-shaped, to 1½ inches long, borne in clusters from the axils of the leaves. Season: May, June.

One of the most interesting habitats for plant life in the northeastern United States is the shale barrens of the central Appalachians. This unusual habitat is found only in a small part of eastern West Virginia and adjacent portions of Virginia, Maryland, and Pennsylvania.

The barrens, named because of the rather sparse vegetation that grows upon them, are developed on slopes where hard, shaley rocks outcrop on the steep hillside. Through frost action, these shaley rocks fracture, and numerous shale fragments are scattered over the slopes.

The nature of the rock fragments and the slopes on which they are strewn prevent the accumulation of much moisture and consequently account for the paucity of plant growth. Little soil is able to develop on the steep slopes, and what is formed usually washes away. Where the loose shale piles up in mounds as it creeps or tumbles downslope, masses of acidic humus accumulate which are too porous to permit a concentration of available nitrogen. It is under these diverse and unique conditions that a peculiar flora has developed on the shale barrens, a flora composed of several species that are found nowhere else in the world. Dr. Earl Core of West Virginia University has reported that there are fifteen different kinds of flowering plants which are found only on these shale barrens of the central Appalachians. These plants are ones which are able to withstand very dry, mineral-poor conditions.

I have often wondered in amazement at how habitats such as deserts and shale barrens and tundra, which seem so uninviting for living organisms, can support vegetation. More remarkable is that there is considerable diversity among the species present in these habitats.

The most famous of the shale barren plants is probably Kates Mountain clover (*Trifolium virginicum*). This species was discovered by John Kunkel Small from Kates Mountain, West Virginia, just outside White Sulphur Springs, on May 16, 1892. Small, who was a botanist for the New York Botanical Garden for many years, was one of the most prodigious writers on the southeastern flora. His *Flora of the Southeastern United States* is still a useful work for the plants of that area. At the time he discovered Kates Mountain clover, Small was only twenty-three years old.

For twenty-four years after its original discovery, *Trifolium virginicum* remained undetected. In May 1916, a second locality was found in West Virginia, this time at Upper Tract in Pendleton County. Then another drought for this species followed, until 1923, when the first Virginia collection was made at Hot Springs.

Many collectors started combing the barrens about 1925, and Kates Mountain clover was collected from a number of sites. The venerable Edgar T. Wherry of Pennsylvania first recorded this clover from Maryland in 1928 and from Pennsylvania in 1932.

Today this shale barren species can be found at a number of localities in West Virginia and Virginia, but it is reduced to two counties in Pennsylvania and to a single locality in Allegheny County, Maryland.

Kates Mountain clover is a perennial whose long taproot penetrates deeply into the shaley substrate. The slender, hairy stems sprawl across the rock-strewn surface. Like all clovers, this one has leaves divided into three leaflets (I've never seen a lucky four-leaf Kates Mountain clover). The individual white flowers are typically pea-shaped and about one-half inch long. They bloom during May and the early part of June.

Although Kates Mountain clover is not yet listed as a federally endangered or threatened species, it is considered vulnerable in West Virginia and a species of special concern in Virginia.

Mountain Pimpernel or *Pseudotaenidia montana.*

Carrot family. Perennial; stems smooth, to 2½ feet. Leaves alternate, compound, divided into 15 or more oblong, toothless leaflets. Flowers very small, yellow, borne in umbrellalike clusters. Season: May.

My first visit to the shale barrens of West Virginia was looked forward to with great anticipation. My appetite was whetted by the possibility of seeing Kates Mountain clover (*Trifolium virginicum*), the white-haired leatherflower (*Clematis albicoma*), Pursh's morning-glory (*Convolvulus purshii*), yellow buckwheat (*Eriogonum allenii*), Harris' goldenrod (*Solidago harrisii*), and the shale evening primrose (*Oenothera argillicola*), all species found at no other place in the world except the shale barrens of West Virginia, Virginia, Maryland, and Pennsylvania.

In Virginia, these barrens are found scattered in the west-central part of the state, in the vicinity of pleasant Virginia villages such as Covington, Hot Springs, Monterey, and Goshen. In Maryland, they are found sporadically west of Hagerstown, between Cumberland and Indian Springs. The Pennsylvania shale barrens are located primarily in the vicinity of Chaneysville and Hewitt in Bedford County.

West Virginia has the greatest number of shale barrens, mostly in two areas of the state. Near the Maryland panhandle and west of Winchester, Virginia, are several barrens between the communities of Keyser and Gore. In the southeastern corner of the state, in the vicinity of White Sulphur Springs, are more occurrences of the barrens.

It was in this latter area in mid-August that I was to study my first shale barrens. The destination was Kates Mountain, where several of these locally occurring plants had been first discovered.

After a sumptuous lunch at the famous Greenbrier Hotel (which I later regretted as I trudged up the western slopes of Kates Mountain), I was eager and ready to see some rare plants.

Once on the barrens, I didn't have to wait long for my first encounter with an endemic. There, in the middle of loose shale, was a prostrate clover, no doubt the long-sought Kates Mountain clover It had bloomed a couple of months earlier, but the very narrow leaflets gave away the identity of the plant. A gorgeous yellow wildflower on a three foot stem was seen a few yards upslope. I knew in an instant it was the shale evening primrose (*Oenothera argillicola*) as I clumsily scrambled, half falling, to the next species as if it were going to get away from me.

All the rest of the plants I had wanted to see, and then some, were there on the flank of Kates Mountain.

Before I left the slopes, however, my eyes fell upon an eighteen-inch-tall plant that was obviously a member of the carrot, or parsley, family because of its compound leaves. The leaves reminded me of the smooth pimpernel (*Taenidia integerrima*), a plant I was familiar with in dry woods in Illinois, but the small fruits with their broad, corky wings indicated to me it was not *Taenidia*. *The Flora of West Virginia*, by Strausbaugh and Core, which I was carrying, directed me to *Pseudotaenidia montana*, the mountain pimpernel.

Pseudotaenidia is a unique genus, comprised only of one species, *P. montana*, which is restricted to the shale barrens of West Virginia, Virginia, Maryland, and Pennsylvania. The plant was discovered by Kenneth K. Mackenzie, the great sedge authority, who found it on Kates Mountain on August 29, 1903. Who knows but what I may have been looking at the same original colony nearly three quarters of a century later. This plant has historically been found several times in West Virginia and Virginia, and it still exists in at least six West Virginia and five Virginia counties. In Pennsylvania and Maryland, the mountain pimpernel has always been very restricted in its distribution. In fact, it hasn't been seen in Pennsylvania since 1936.

Pseudotaenidia montana is completely smooth on all its parts. Its yellow flowers are borne during May in a compound umbel that surmounts the stem.

The restricted habitat and relatively few populations of this species demand that it be monitored continuously. It is considered rare and vulnerable in West Virginia and a species of special concern in Virginia.

Heart-Leaved Plantain or *Plantago cordata*.
Plantain family. Perennial; leaves basal, heart-shaped, to 8 inches long, to 6 inches broad. Flowers crowded in slender, elongated clusters on leafless stalks to 18 inches; petals 4, nearly transparent. Season: March, April.

The name plantain to many people brings to mind herbicides, digging tools, and other devices one uses to rid lawns of coarse, aggressive weeds. Common plantain (*Plantago rugelii*) and buckhorn plantain (*P. lanceolata*) have plagued homeowners for many years. Although these two European imports which flower throughout the summer have given plantains a bad name, there are several truly American plantains which live in pristine native habitats.

One of these natives is *Plantago cordata*, the heart-leaved plantain. This plant is always associated with heavily shaded streams. Sometimes the streams are gravel-bottomed. I have watched one colony of this plantain for years in southern Illinois, where it grows in clear, flowing water one or two inches deep.

During the last part of the nineteenth century and the first three decades of this century, the heart-leaved plantain grew from New York to Minnesota in the north and Alabama to Louisiana in the south. It also occurred occasionally in southern Ontario. But after 1935 the plant began to decline drastically. When M. F. Tessene, a student at the University of Michigan, studied *Plantago cordata* in 1968, he was able to find only one colony in Ontario, two in New York, one in North Carolina, two in Illinois, and one in Missouri. Botanists began predicting that this species would become extinct. This was further amplified when it was discovered that the seeds of this species remained viable for only a very short time, and that most of the seeds never sprouted.

During the 1970s botanists made a concentrated effort to locate additional colonies of the heart-leaved plantain. There was considerable success in Missouri, where this species was found in several clear streams. In New York this plantain was discovered along a 120-mile stretch of the Hudson River. In other parts of its range, however, *Plantago cordata* is rare and apparently continues to decline in numbers.

This perennial has a tuft of deep green, heart-shaped leaves that may reach a width of six inches. From among these leaves arises

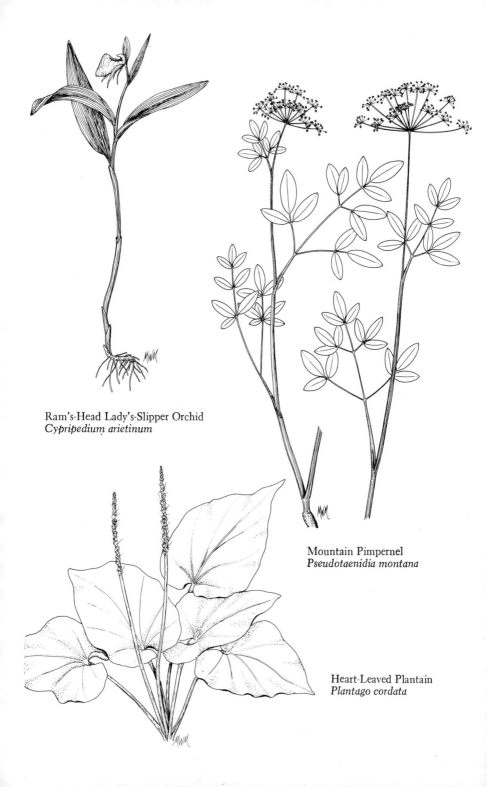

Ram's-Head Lady's-Slipper Orchid
Cypripedium arietinum

Mountain Pimpernel
Pseudotaenidia montana

Heart-Leaved Plantain
Plantago cordata

a slender, leafless flower stalk that bears small, inconspicuous flowers near the tip. The flowers bloom in March and April.

Because of the rarity of this species, it is under review for possible listing as a federally endangered species.

Ram's-Head Lady's-Slipper Orchid or *Cypripedium arietinum*.

Orchid family. Perennial; stems hairy, to 12 inches. Leaves 3–4, elliptic, pointed at the tip, to 4 inches long. Flower solitary at tip of stem; sepals 3, purple to green, to 1 inch; lip petal white with purple veins, swollen. Season: May, June.

Lady's-slippers receive their name from the enlarged lip petal which has a fanciful resemblance to Cinderella's slipper. But in one lady's-slipper the lip petal has a pouch which projects downward, rather than being slipper-shaped. Some of the early botanists compared the general shape of the lip to that of a ram's head, and the orchid gradually became known as the ram's-head lady's-slipper, *Cypripedium arietinum*.

This orchid occurs in a wide band across southern Canada, from Saskatchewan to Quebec. In the United States it is confined to western Maine, Vermont, New Hampshire, western Massachusetts, northern Connecticut, eastern New York, northern Michigan, northern Wisconsin, and northeastern Minnesota. It is uncommon to rare in each of these states.

New York botanists note that the ram's-head lady's-slipper occupies two types of habitats. One is an open forest dominated mostly by cone-bearing trees; the other is a swamp forest. When I was studying the rare plants of Michigan's Upper Peninsula, I observed *Cypripedium arietinum* in conifer swamps dominated by black spruce and tamarack and in cedar swamps dominated by white cedar, balsam fir, white birch, American elm, and black ash.

The ram's-head lady's-slipper is a rather inconspicuous plant with slender stems that grow no more than one foot tall. Three or four narrow, pointed, bluish-green leaves are found along the stem. From the tip of the stem is an inch-long purple and white flower with the characteristic ram's-head lip petal. The flowers bloom in May and June.

The rarity of this species in the United States makes it of special concern for botanists in several states.

3

Dwindling Species
of the Southeastern States

Botanists in the southeastern United States have been very forceful in justifying several of the plants in that area for federal listing. The twenty plants discussed in this chapter include nine which already are listed as either federally endangered or threatened. The remaining eleven are rare species that are under review for possible future listing.

States considered southeastern in this book are Virginia, Kentucky, Tennessee, North Carolina, South Carolina, Georgia, Alabama, Florida, and Mississippi.

Virginia Round-Leaf Birch or *Betula uber.*
Birch family. Tree to 35 feet. Leaves alternate, simple, nearly round, heart-shaped at base, coarsely toothed, to 2 inches. Flowers inconspicuous, without petals, borne in spikes. Season: April, May.

Among the more than 3,000 kinds of plants listed by the Smithsonian Institution in its report to Congress in 1975 were 100 which were thought to be extinct in the United States. One of these was a small birch tree from Smythe County, Virginia, which had not been seen for nearly sixty years.

It was in June 1914 when William W. Ashe, a prominent botanist with the United States Forest Service, collected specimens of an unusual birch along the banks of Dickey Creek south of Sugar Grove Station, Virginia, at an altitude of 2,800 feet. The strange birch was found growing among specimens of sweet birch (*Betula lenta*). Both birches had the characteristic wintergreen flavor in their bark, and the only recognizable difference between the two was the shape of the leaves. Sweet birch has leaves which are sharply pointed at the tip, while the new birch's leaves are rounded at the tip.

Four years after his discovery, Ashe described his birch as a variety of *Betula lenta*, calling it *Betula lenta uber*. For forty years this plant was apparently forgotten. Then, in 1945, Merritt Lyndon Fernald of Harvard University reopened the saga of the Virginia round-leaf birch and came to the conclusion that it deserved to be considered a species. He called it *Betula uber*. He based his conclusions on the leaf shape and on characters of the female flowers. (The male flowers were not known.)

After Fernald called attention to this plant, a number of people tried to relocate it along Dickey Creek, but without success. Some botanists speculated that it may have been an abnormal specimen of *Betula lenta* that Ashe just happened upon.

Enter Peter Mazzeo into the picture in 1970. Mazzeo, a research botanist at the United States Arboretum in Washington, D.C., was working on the flora of Virginia and was preparing a treatment on the birches. Through diligent research, Mazzeo discovered that additional specimens of this round-leaf birch were filed away in the herbarium of the Arnold Arboretum. These had been found by a private forester, H. B. Ayres, from the banks of Cressy Creek, Virginia. A check in the atlas showed that Cressy Creek was just a little east of Dickey Creek. Mazzeo published the findings of his research in a quarterly Virginia botanical newsletter called *Jeffersonia*, indicating that Ayres had discovered *Betula uber* along Cressy Creek.

Spurred on by Mazzeo's article, Douglas Ogle, a biology teacher at Virginia Highlands Community College in Abingdon, decided to spend the summer of 1975 searching for the Virginia round-leaf birch. On August 22, while exploring along the west bank of Cressy Creek, Ogle rediscovered the elusive birch.

In mid-September, Ogle, Mazzeo, other botanists and interested observers, and the private landowners where the birch was found made an excursion along Cressy Creek. They found twelve mature trees, one of which was thirty-five feet tall, and sixteen seedlings. Several of the seedlings had been damaged by cattle which were grazing the area. In addition, they found male flowers, which had never been observed before, as well as female flowers developing for next spring.

From 1975 to 1978 additional searchers turned up a total of forty individuals of the Virginia round-leaf birch. Of these, fourteen were mature plants and the remainder seedlings and saplings. Three of the individuals were found in the Jefferson National Forest adjacent to the two private parcels of land where the others grew.

Because of its great rarity and limited range, the Virginia round-leaf birch was placed on the federally endangered species list on April 26, 1978.

By July 1980 only twenty of the forty individual plants had survived. Four of them died from apparent natural causes, while eight of the seedlings were transplanted to various gardens or research areas to ensure the species' existence. Another eight seedlings were presumably transplanted, but there is no documentation for this.

High fences have been erected around all the known specimens, and a *Betula uber* Protection, Management, and Research Coordinating Committee was set up. Attempts were made to propagate the Virginia round-leaf birch. Fifty individuals were produced from rooted cuttings at the National Arboretum, but grafting experiments onto other birch rootstocks failed.

In accordance with provisions in the Endangered Species Act, a team of biologists was organized by the United States Fish and Wildlife Service to prepare a recovery plan for the Virginia round-leaf birch. The recovery plan was approved March 3, 1982. It gives a step-by-step plan on how to preserve the existing specimens and how to increase their numbers.

The Virginia round-leaf birch has relatively smooth, dark bark. The leaves, often nearly circular in outline, are heart-shaped at the base and coarsely toothed along the margins. The flowers are borne in spikes, known as catkins, which develop during the autumn in preparation for their maturing by the next spring.

Harper's Beauty or *Harperocallis flava*.
Lily family. Perennial. Leaves basal, slender, stiff. Flower solitary on a nearly leafless, 2½-foot stalk, yellow; petals ½ inch long. Season: May.

One might think that since John Clayton and Peter Kalm first collected plants in the United States about 250 years ago that all the different kinds of flowering plants in the country would have been discovered and classified by now. Such is not the case. New species of plants are being discovered every year in this nation, mostly in the western states, where vast areas still remain essentially unexplored and certainly unbotanized. It is unusual, however, to discover a new genus for this country, particularly east of the Mississippi River, where our knowledge of the flora has been well documented.

But Prof. Sydney McDaniel of Mississippi State University did just that in 1965. On May 4 of that year, while exploring in the Florida panhandle in Franklin County, he observed a yellow-flowered member of the lily family which turned out to be not only a new species but also a new genus.

McDaniel named his remarkable discovery *Harperocallis flava*. *Harperocallis* was in honor of Roland M. Harper, a noted plant explorer and botanist of the southeastern United States during the first half of the twentieth century. (Ironically, Harper died just two years before the genus was named for him.)

Although this new species is not as showy as a tiger lily or a turk's-cap lily or other colorful members of the lily family, it is attractive enough to have earned the common name of Harper's beauty.

Harper's beauty is a perennial herb that grows from an underground rhizome. Several stiff, slender leaves arise from near the base of the plant. In April, a slender, unbranched stalk emerges from among the leaves. Only one yellow flower is formed at the tip of the stalk and, although relatively small by lily standards, it is bright and showy. It usually is open by early May.

McDaniel discovered his new species in flower in May, and then returned to the same locality in September and found it with capsules and seeds. Dr. Robert K. Godfrey, premier botanist at Florida State University, gathered specimens of Harper's beauty from Franklin County in 1969. All of the plants that McDaniel and Godfrey found were less than half a mile from each other.

Then, on April 28, 1975, Sydney Thompson found a new colony of Harper's beauty in Liberty County, nineteen miles north of the original site. All known localities are within the Apalachicola National Forest.

The original area (called the type locality) where *Harperocallis flava* occurs is in an open bog surrounded by the buckwheat tree

(*Cliftonia monophylla*) and the odorless bayberry (*Myrica ino-dora*). Nearby are stands of longleaf, slash, and pond pines (*Pinus palustris, P. elliottii,* and *P. serotina*).

The Apalachicola River region is known for its endemic species, that is, species found at no other place in the world. Nine other endemics live here: the Florida torreya (*Torreya taxifolia*), Florida yew (*Taxus floridana*), wiregrass gentian (*Gentiana pennelliana*), mock pennyroyal (*Hedeoma graveolens*), Godfrey's blazing-star (*Liatris provincialis*), white bird's-in-a-nest (*Macbridea alba*), Ashe's magnolia (*Magnolia ashei*), Apalachicola rosemary (*Conradina glabra*), and Chapman's crownbeard (*Verbesina chapmanii*).

Each of these ten is considered either endangered or threatened in the state of Florida, while Harper's beauty is listed as federally endangered (as of October 2, 1979).

Mountain Golden Heather or *Hudsonia montana.*
Rockrose family. Dwarf shrub; stems to 6 inches. Leaves short, needlelike. Flower solitary at end of leafy shoot, yellow, to 1 inch across; petals 5. Season: June, July. (See color plate 3.)

The Linville Gorge area of the Pisgah National Forest in the Blue Ridge Mountains of western North Carolina provides for some of the most spectacular mountain scenery in the eastern United States. Deep wooded ravines contrast with bare granitic ledges; clear mountain streams penetrate the area.

It is on a few of the exposed ledges that an attractive but very rare species of flowering plants occurs. The plant is known as the mountain golden heather (*Hudsonia montana*), a member of the Cistaceae, the rock rose family. While this species is a mountain plant, the other species of *Hudsonia* in the eastern United States occur on dry sands or on sand dunes. In addition, there are no other golden heathers within at least one hundred miles of the mountain golden heather.

Thomas Nuttall was the first person to collect this species when he made an extensive foray in 1816 across the mountains of North Carolina. Nuttall, who was born in Yorkshire, England, came to the United States in 1808 when he was twenty-two years old and immediately began his career as a botanist. Between 1809 and 1818, he collected extensively in the United States. From his collections he published an important two-volume work in 1818, entitled *The Genera of North American Plants,* in which he named *Hudsonia montana.*

A few botanists who ventured into the Linville Gorge after Nuttall's early trip saw the mountain golden heather and made occasional collections of it. When Nuttall first described this species, he noted that it was "abundant," but later collectors reported it as "not plentiful" or "scarce." Botanists who tried to find it during the late 1960s couldn't find it at all, and there was speculation that the mountain golden heather had become extinct.

In recent years, Larry Morse of the New York Botanical Garden and Ben Sanders of the United States Forest Service have searched for *Hudsonia montana* and found it at four locations, all within a radius of five miles.

The habitat for the mountain golden heather is on nearly bare, exposed quartzite ledges. Except for a scattering of mosses and lichens on the rock surface, most of the vegetation associated with this plant are shrubby members of the heath family, particularly the sand myrtle (*Leiophyllum buxifolium*). These heath-dominated, open, rocky areas are referred to as heath balds.

Despite the fact that the mountain golden heather lives on high, rocky ledges, it is in some danger from intensive recreation activity in the Linville Gorge area because of the popularity of the region. At one site, a camper built a campfire directly on a small colony of this species. Excessive trampling by hikers and rock climbers poses some threat to it. Several plants may grow clumped together, forming a spreading colony. Because the leaves are short and needlelike, the colony might be mistaken for a low-growing juniper when it is not in flower. A solitary bright yellow flower nearly an inch across appears during June at the end of each leafy shoot.

Although only four locations are known for this species, it is thought that perhaps 2,000 individuals exist. Because of this, the mountain golden heather was listed on October 20, 1980, as threatened on the federal roster.

Bunched Arrowhead or *Sagittaria fasciculata.*
Water plantain family. Perennial. Leaves spatula-shaped, to 1 foot. Flowers several on a leafless stalk; sepals 3, green; petals 3, white. Season: May, June. (See color plate 4.)

It may be romantic to think of an endangered species occurring on a remote mountaintop, or on a wind-swept island, or in a majestic virgin forest, but rare plants may be found in any kind of habitat, including some which have been degraded by man's activities.

Such is the case of the bunched arrowhead (*Sagittaria fasciculata*). The bunched arrowhead has never been known from more than three areas in North Carolina and one in South Carolina, and two of the North Carolina sites no longer support this plant. The two remaining locations are in areas of high disturbance, and the future of the bunched arrowhead is much in jeopardy.

At the North Carolina site in the vicinity of Flat Rock, the bunched arrowhead occurs in a creek which is adjacent to a well-traveled highway and a major railroad. Railroad maintenance work in 1979 severely damaged the existing population, and annual use of herbicides by the railroad to keep down the weeds probably adversely affects the species. In addition, there is concern that any widening of the highway might completely obliterate the Flat Rock population.

In South Carolina, where the bunched arrowhead grows in the vicinity of the Enoree River, it occurs in a power-line right-of-way. It is easy to imagine that use of the right-of-way to maintain and service the power lines could destroy the South Carolina plants overnight.

Because of the precarious situations surrounding the bunched arrowhead, it was officially designated as a federally endangered species on July 25, 1979.

The first collection apparently ever made of this species was in May 1887 from the Flat Rock area, but botanists failed to recognize the plant as a new species and erroneously called it *Sagittaria macrocarpa*, which is a different species. It was not until the late Ernest O. Beal, a professor of botany and wetland plant specialist at Western Kentucky University, made a thorough study of the arrowheads of the Carolinas in 1960 that *Sagittaria fasciculata* was recognized as a new species.

Details surrounding one of the locations of the bunched arrowhead are not precise. A specimen was collected on May 26, 1896, "in sluggish streams, Biltmore, North Carolina." There is no indication who the collector was, but the specimen was included in the historic herbarium maintained at Biltmore. Botanists for nearly a century have combed the sluggish streams near Biltmore, trying to relocate this site, but have met with no success.

For a while, a colony of bunched arrowhead occurred along a railroad north of Hendersonville, North Carolina, but this station apparently has been destroyed.

Only in recent years has the bunched arrowhead been found in South Carolina. It was found by Dr. Leland Rodgers of Furman University.

The bunched arrowhead is a perennial which grows in standing, but not stagnant, water. Unlike most arrowheads, which are often coarse, this species rarely grows taller than one foot. The leaves are not arrowhead-shaped, either, but are described as spatula-shaped.

One or more leafless flower stalks, originating from the base of the plant, bear a few clusters of white flowers in May and June.

Persistent Trillium or *Trillium persistens.*
Lily family. Perennial; stem smooth, to 10 inches. Leaves 3 in a ring on the stem, lance-shaped to ovate, to 4 inches. Flower solitary, drooping; sepals 3, green; petals 3, white, to 1 inch. Season: March, April. (See color plate 5.)

There are two things that impressed me about the magnificent Tallulah Gorge in northeastern Georgia. One is the 900-foot-deep gulf that The Great Karl Wallenda tightroped across in 1970, and the other is the fact that it is the home for the endangered persistent trillium (*Trillium persistens*).

Tallulah Gorge is surrounded by the very scenic and vegetation-rich Chattahoochee National Forest. Waterfalls abound in this national forest, the most spectacular being Anna Ruby Falls, located high on the forested slopes of Tray Mountain. Virgin forest conditions prevail at Sosebee Cove, where majestic tulip poplars and huge Ohio buckeyes tower over a rich layer of ferns and wildflowers. The Appalachian Trail has its southern terminus in the Chattahoochee.

Wilbur Duncan of the University of Georgia was collecting plants in Tallulah Gorge on July 6, 1950, when he found a trillium in fruit which looked strange to him. Unable to identify the plant, Duncan set it aside for later study.

Twenty years later, Edna Garst of Athens, Georgia, found a

flowering trillium just across the state line into South Carolina, about four miles from Tallulah Gorge. When Mrs. Garst's plant was compared with the fruiting trillium that Duncan had collected in 1950, the two appeared to represent the same species, and one that seemingly was new to science.

Spurred on by this discovery, Drs. John Garst (Edna's husband) and George Neese, chemists from the University of Georgia, searched the area around Tallulah Gorge and found several hundred flowering specimens of this trillium.

There was enough evidence now to conclude that the plants belonged to a new species, which Duncan named *Trillium persistens* in 1971. The epithet *persistens* is derived from the fact that this species remains green long after all the other trilliums of the forest have withered. (It also blooms about one month earlier than most of the others.)

The persistent trillium occurs as scattered individuals in the forest, never growing in groups of more than ten plants.

The genus *Trillium*, a member of the lily family, is named for several parts of the plants which are in threes. There are three leaves on each stem, and the flower has three sepals, three petals, six stamens, a three-parted style, and a three-lobed ovary. Several species of trillium occur in the eastern United States, where they form a conspicuous element of the spring flora. Most of them are exceptionally attractive, and several are widespread and common.

The smooth stems of *Trillium persistens* stand from five to ten inches tall. A single white flower is formed on a one- or two-inch-long stalk that emerges from each cluster of leaves. Usually the stalk is arching and the flower more or less drooping. As the petals age, they tend to turn reddish purple, except for the base, which remains white. Flowering occurs in late March and early April.

Trillium persistens is confined to a four-mile area in Rabun and Habersham counties, Georgia, and Oconee County, South Carolina. At every place where it has been found, the persistent trillium grows beneath two species of rhododendron (*Rhododendron maximum* and *R. minus*) in decomposed litter. Several other trilliums usually occur in the same woods.

The very limited range of this species enabled it to be listed as federally endangered on April 26, 1978.

Hairy Rattleweed or *Baptisia arachnifera.*

Pea family. Perennial; stems to 2 feet, densely hairy. Leaves alternate, heart-shaped, densely hairy, to 3 inches. Flowers yellow, sweet-pea-shaped, in elongated clusters. Season: June, July.

Baptisia is a generally well known genus of perennial, often shrub-like herbs belonging to the pea family. Gardeners cultivate several kinds because of their handsome sprays of lupinelike flowers. The blue false indigo (*Baptisia australis*) is probably the favorite of the garden types, but the bright yellow rattleweed (*B. tinctoria*) and the white-flowered *B. leucophaea* are also popular.

Craftsmen working with natural dyes sometimes use *Baptisia tinctoria* for a source of indigo, although traditionally species of the genus *Indigofera* were the major suppliers of the natural indigo. Because of their secondary importance as a producer of indigo, Baptisias are usually called false indigos.

Prairie enthusiasts are thrilled by the presence of another *Baptisia, B. leucantha,* the lead plant, because they know that this species is a good indicator of a relatively undisturbed prairie.

Dr. Wilbur Duncan, now emeritus professor of botany at the University of Georgia, happened upon an unusual-looking *Baptisia* while specimen hunting in the vicinity of Jessup, Georgia. He initially discovered this silvery-leaved plant when it was bearing its yellow flowers. Returning to the site in September, he was able to obtain the legumes from the plants.

Sometimes botanists have difficulty trying to decide whether a plant which is somewhat different from its relatives is really a new species or merely a variation of an already known species. But Dr. Duncan had no trouble determining that what he had found was unlike any other *Baptisia* in the world.

Thus, in 1944, Duncan named this new species *Baptisia arachnifera,* referring to the arachnoid, or cobwebby, hairs found all over the stems and leaves of the plant.

During the time since its discovery, *Baptisia arachnifera,* which has become known as the hairy rattleweed, has been found only in an area about ten miles long in Wayne and Brantley counties, Georgia. It occurs in pine-dominated flat woods or on low sandy rises. Slash pine and longleaf pine dominate these woods, with the

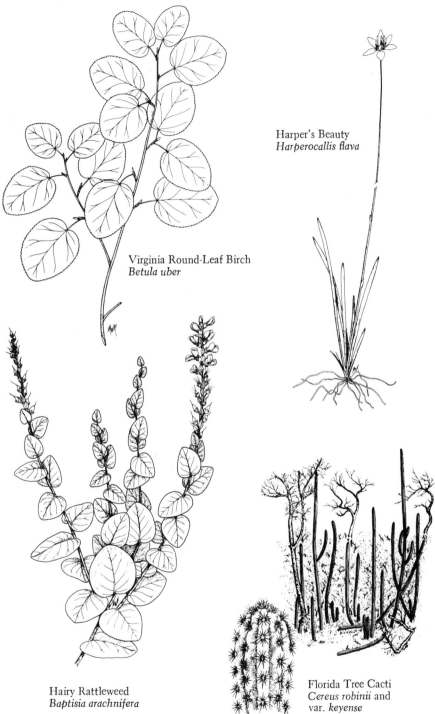

Virginia Round-Leaf Birch
Betula uber

Harper's Beauty
Harperocallis flava

Hairy Rattleweed
Baptisia arachnifera

Florida Tree Cacti
Cereus robinii and
var. *keyense*

saw palmetto and gallberry being the characteristic shrubs. It appears to prosper in areas where there has been recent fire.

The hairy rattleweed will seed into areas which have been prepared for pine, and it will occur as an understory plant in young plantations. However, as the pines mature and the shade on the forest floor increases, the hairy rattleweed disappears.

Baptisia arachnifera is a perennial herb with a rather stout, much branched stem growing to a height of about two feet. Both the stems and the leaves are covered with a dense mat of cobwebby-gray hairs. Arising from the axils of the uppermost leaves are six-inch-long clusters of yellow, sweet-pea-shaped flowers. The fruit is a hairy legume with a short beak at the upper end.

The main threat to the hairy rattleweed is that lumber companies working the area where this plant occurs might modify the habitat in such a way that the species will perish.

Baptisia arachnifera was listed as a federally endangered species on April 26, 1978.

Chapman's Rhododendron or *Rhododendron chapmanii.*

Heath family. Evergreen shrub to 6 feet. Leaves alternate, simple, elliptic, without teeth, to 3 inches. Flowers in short, compact, showy clusters, rose-pink; petals united, 5-parted, funnel-shaped. Season: March, April. (See color plate 6.)

Among the most beautiful shrubs in the world are rhododendrons and azaleas. Their huge clusters of bright flowers, often set against a background of glossy green leaves, are some of nature's most elegant creations.

Most botanists today consider both rhododendrons and azaleas to belong to the same genus, *Rhododendron.* Rhododendrons generally keep their leaves throughout the winter months, while azaleas are deciduous.

There are more than 600 species of rhododendron in the world, and countless cultivated varieties and hybrids. The wild species center on three regions. Many of them are native to the Far East, some are in the northwestern United States, while several grow wild in the southeastern part of the country.

Rhododendron chapmanii, a plant of the southeastern United States, is probably the rarest of all the North American species in the genus. The name for this plant commemorates its discoverer, Alvan Wentworth Chapman. Chapman found it during the 1800s in the Florida panhandle, where it was growing in a sandy pine barren.

Although Chapman's profession was that of a physician, practicing most of his life in Quincy, Marianna, and Apalachicola, Florida, he spent all of his spare moments plant hunting in northern Florida and adjacent areas. His botanical work climaxed in the publication of *Flora of the Southern United States* in 1860.

Chapman's rhododendron has been found in five counties of the Florida panhandle and in Clay County in northeastern Florida, not more than forty miles from the Atlantic Ocean. There are unverified reports of this species from southern Alabama and southern Georgia.

The usual habitat for this species is a pine woods in the transition zone between dry forests of longleaf pine and turkey oak and moist thickets dominated by the titi bush (*Cyrilla racemiflora*). However, in Clay County, where its associates are sand pine, saw palmetto, twin oak, Chapman's oak, American holly, and two kinds of stagger-bushes, Chapman's rhododendron lives on a sandy bluff along a branch of a creek.

One of the panhandle stations was clear-cut in 1970, and much of the existing population was destroyed. Nonetheless, several plants were observed sprouting in the cleared area in 1977 after the area had been burned and cultivated prior to replanting to pine. There now are about 500 Chapman's rhododendrons at this site.

Only three localities are known definitely for this attractive shrub today. Because of the very reduced number of specimens still living at these sites, Chapman's rhododendron was listed as a federally endangered species on April 24, 1979.

Rhododendron chapmanii is an evergreen shrub with its many stems reaching a height of only about six feet. The rose-colored flowers, which are borne in short, compact clusters, open in March and April.

Although this species has probably never been very common, its recent decline has been due to clear-cutting the habitat for paper production, digging clumps of the species for the nursery trade, and draining of titi bogs for subsequent pine plantings.

Tennessee Coneflower or *Echinacea tennesseensis.*
Aster family. Perennial; stems bristly hairy, to 1 foot. Leaves alternate, simple, toothless, densely hairy, to 6 inches. Flowers crowded together in a solitary head with 12–15 spreading, magenta rays surrounding a darker, central disk of tubular flowers. Season: July. (See color plate 7.)

It was early afternoon on a mid-July day when my wife, my seventeen-year-old son Trent, and I arrived in downtown Nashville, Tennessee, and headed through the historic section of the city to the Old Customs Building at Seventh and Broadway. This historic building contains the offices of the Tennessee Heritage Program, where Dr. Paul Somers, an authority on the endangered Tennessee coneflower (*Echinacea tennesseensis*), is employed.

The Heritage Program is one of several operating in the United States in which The Nature Conservancy and state governments work hand in hand to create a continuing process for identifying significant natural areas and setting priorities for protecting them. A private organization, The Nature Conservancy is known for having purchased many valuable parcels of land containing pristine natural areas in all parts of the United States. Their innovative Heritage Programs began in 1974 and today are operating in more than half of the states in the country.

Dr. Somers has spent the last few years observing the four known localities in the world for the Tennessee coneflower, and he agreed to take me to one of the sites near Nashville. All four locations occur within fourteen miles of each other on limestone glades. The glades, which are frequent in middle Tennessee and adjacent Kentucky and Alabama, have long been known for their unique flora.

We proceeded into a semirural area where Paul motioned for me to pull over to the shoulder. With the sounds of the city audible in the distance, we walked across low vegetation which was crowding the road, and through an opening in a row of spindly trees. Fifteen feet from the van, we were in a glade. Small flat slabs of slate-gray limestone lay strewn across the flat terrain. A great variety of herbs, most of them under fifteen inches tall, was scattered over the glade. Only an occasional red cedar and Carolina buckthorn interrupted the pattern. The herbaceous plants were a fascinating mixture of species. Some are plants that are restricted to the glades, such as the prostrate Gattinger's prairie clover (*Petalostemum gattingeri*), and the exceedingly uncommon Eggleston's violet (*Viola egglestonii*) and Tennessee milk vetch (*Astragalus tennesseensis*). Others are those species tolerant of dry, rocky openings in general. These include the prickly pear cactus (*Opuntia compressa*), slender yellow flax (*Linum sulcatum*), and rushfoil (*Crotonopsis elliptica*). Here and there the fleshy basal leaves of the American agave (*Polianthes virginica*) were clustered about a spent flower stalk.

At the far end of the glade was the Tennessee coneflower. Dozens of magenta heads, all facing the same way, greeted us. We hastened to the patch, my son Trent accidentally flushing a camouflaged nighthawk nesting among some nearby gravel. The bird gave us the hurt-wing routine to distract us from the solitary egg she had been protecting.

The bristly hairy stems of the coneflower stand about one foot tall. Terminating each stem is a showy head with about twelve to fifteen spreading, inch-long rays. A dark, central area of short, tubular flowers completes each head.

There were at least six locations for the Tennessee coneflower in the past, but two of them no longer exist. The four remaining populations are subjected to a number of potential threats, which include grazing and housing development. One site, on industrial property, could be eliminated by expansion of the industry.

In an effort to offer some protection to the remaining Tennessee coneflowers, the United States Fish and Wildlife Service listed it as a federally endangered species on June 6, 1979.

Green Pitcher Plant or *Sarracenia oreophila*.

Pitcher plant family. Perennial. Most leaves modified into pitchers, green or yellow-green, to 2 feet, with a hood; a few leaves not pitcherlike, narrow, to 6 inches. Flower solitary on a 2-foot leafless stalk, yellow. Season: April to June. (See color plate 8).

The wonderful world of flowering plants, with its more than 225,000 species throughout the universe, is an array of fantastic and unbelievable forms. High on the list of those plants which have adapted to bizarre and unusual life-styles are the so-called carnivorous, or insectiverous, plants.

Sundews, bladderworts, pinguiculas, the Venus fly trap, and pitcher plants are all groups which have developed ways to entrap animal life and then use parts of their prey for their own nutrition.

In pitcher plants, some of the leaves are modified into elongated tubes, the pitchers, which collect liquid. At one side of the top of each pitcher is an arching hood, which often has a vivid color pattern. Insects of various sorts, in search of nectar and attracted by the colorful hoods, may wander onto a slippery area of the pitcher just below the attachment of the hood. Unable to avoid slipping, the insect falls into the liquid and perishes. It is unable to climb out of the pitcher because of downward-pointing hairs on the inner face of the tube. Digestive enzymes within the liquid enable the pitcher plant to obtain nutrients from the dead organisms.

North American pitcher plants, which are assigned to the family Sarraceniaceae, are grouped into two genera. One of these, the cobra plant (*Darlingtonia californica*), which is confined to Oregon and northern California, is discussed in Chapter 7. The other genus is *Sarracenia*, the pitcher plant, which has ten species in the eastern United States.

Because of their unusual features, as well as their general attractiveness, pitcher plants have been sought by horticulturists and other plant enthusiasts for many years. This has resulted in a drastic decline in the numbers of some of the species in the wild.

One *Sarracenia* whose survival is a matter of concern is the green pitcher plant, *Sarracenia oreophila*. This species occurs today in a small area of northeastern Alabama and a single site in northern Georgia.

Charles Mohr, who was Alabama's pioneer botanist, mentioned as early as 1897 an unusual pitcher plant that was growing along Little River in the mountains of DeKalb County, Alabama. Although Mohr failed to give his discovery a new name, Thomas Kearney in 1900 decided it was a different variety of the more common yellow pitcher plant and called it *Sarracenia flava* var. *oreophila*.

Edgar T. Wherry of the University of Pennsylvania elevated this plant to the rank of a full species when he called it *Sarracenia oreophila* in 1933.

Through the years, this pitcher plant has been found at several localities in Jackson and DeKalb counties, Alabama, at one area in

Cherokee County, and apparently in Etowah County. Plants known to occur in Marshall County, Alabama, may not be a native population.

In Georgia, a specimen of the green pitcher plant was made many years ago from Taylor County in the west-central part of the state, but the plant apparently no longer occurs there. It wasn't until the late 1970s that *Sarracenia oreophila* was verified as growing in Georgia, when W. Michael Dennis of the Tennessee Valley Authority at Muscle Shoals discovered it in the Blue Ridge Mountains of Towne County.

There is an interesting report of the green pitcher plant from Tennessee. A nonflowering *Sarracenia* appearing to be *S. oreophila* was found in a bog in Fentress County in 1935. It was dug and taken to the University of Tennessee greenhouse, where it was potted with the hope that it would live and produce flowers to verify its identity. But the pot was accidentally knocked to the floor and broken. Before the pitcher plant could be saved, a student greenhouse worker had discarded plant, pot, and soil. No one has been able to find a Tennessee specimen of green pitcher plant since.

Sarracenia oreophila is more of a mountain species than any of the other Sarracenias. Its primary habitats are on shady stream edges and in moist woodlands. However, at the Towne County, Georgia, site, the green pitcher plant grows in an open, cleared habitat which is subjected to grazing. The soil is highly organic and is underlain by granite, which becomes exposed in places. It is reported in the March 1977 *Carnivorous Plant Newsletter* that one colony in northeastern Alabama occurs in a hard clay soil containing very fine silica sand and is badly overgrown by shrubs.

Sarracenia oreophila is a perennial herb with underground rhizomes. The pitcher-shaped leaves may reach a length of more than two feet. The hood, which is less than three inches long, may be yellow-green throughout, or it may be purple-spotted at its base, or it may have a conspicuous network of purple veins.

A single yellow flower terminates the often two-foot-long leafless flower stalk. It opens from mid-April to early June.

Besides the threat from commercial collectors, the green pitcher plant is subject to habitat destruction due to urban expansion, strip-mining for coal, and road construction.

Sarracenia oreophila was declared a federally endangered species on December 19, 1979.

Canebrake Pitcher Plant or *Sarracenia alabamensis.*

Pitcher plant family. Perennial. Leaves yellow-green with red veins, pitcher-shaped, to 18 inches, with deep red hoods. Flower solitary on a 2-foot leafless stalk, maroon. Season: April to June. (See color plate 9.)

Perhaps as rare as the green pitcher plant, if not rarer, is the canebrake pitcher plant (*Sarracenia alabamensis*). Although not yet recognized as a federally endangered species, it is listed in the *IUCN Plant Red Data Book*. This publication of the International Union for Conservation of Nature and Natural Resources lists selected species of plants from all over the world which are in danger of extinction.

The canebrake pitcher plant is limited to a very few areas in the central Alabama counties of Autauga, Chilton, and Elmore. Frederick Case in 1974 described the habitat as "somewhat sandy-gravelly bogs, usually on sloping ground, or in damp peaty or mucky soil around small springheads and tiny rills. It grows most often in . . . the 'canebrakes' of the region, or in open boggy swales." Canebrakes are dense, often impenetrable thickets of the woody grass known as the giant cane.

It is estimated that there are probably fewer than 500 of these pitcher plants remaining in the wild. They have been subjected to intensive commercial digging. In the Chilton County *News* in 1975, one collector ran an advertisement offering a twenty-dollar reward for verified locations for the canebrake pitcher plant.

Some of the areas where this species occurs are disturbed and subject to encroachment by alien plants. Dr. George Folkerts of Auburn University, an authority on pitcher plants, has described one site as overgrown with honeysuckle and damaged from herbicide spraying.

Sarracenia alabamensis is a member of the red pitcher plant (*Sarracenia rubra*) complex. A number of botanists continue to call the canebrake pitcher plant *Sarracenia rubra* ssp. *alabamensis,* since it shows some intergradation with *Sarracenia rubra*. A few persons think it may only be a shade form of the red pitcher plant.

Frederick Case has given fairly good evidence that the plant should be called a distinct species. It is a rhizome-forming perennial with a cluster of hooded pitchers at ground level. The pitchers range from eight to eighteen inches tall and are rather variable in coloration. They normally are yellow-green with a conspicuous network of bright red veins, particularly on the inner surface near the top.

The hood strongly arches over the opening of the pitcher and frequently is deep red, particularly in late summer and autumn. Maroon-colored flowers are borne singly at the ends of eighteen-inch-long stalks.

Price's Groundnut or *Apios priceana.*

Pea family. Vine. Leaves alternate, compound, divided into 3–9 ovate leaflets to 4 inches. Flowers rose-pink, sweet-pea-shaped, to 1 inch, with a swollen, spongy tip, borne several in a cluster. Pods to 6 inches. Season: July. (See color plate 10.)

It must have taken much determination and dedication for the naturalists of the nineteenth century to conquer the wilderness and make collections of plants which were to establish the foundation on which our modern floras are based. Not only was it often rugged blazing new trails, but it was also very difficult at times to identify the specimens found because of the scarcity of floras.

It is even more remarkable that one of the pioneer botanists to collect in the hinterland of Kentucky was Sarah Francis Price. Sadie, as she was known, was the discoverer of several species new to science, including Price's sorrel (*Oxalis priceae*), Price's dogwood (*Cornus priceae*), and the subject of this account, Price's groundnut (*Apios priceana*).

Sadie Price was born in 1849 and attended St. Agnes school for girls in Terre Haute, Indiana, where she developed a strong interest in biology and art. After her schooling, she returned to Bowling Green, Kentucky, where she lived with her sister until Sadie's unexpected death in 1903. Between her schooling and the time she began teaching school, she became afflicted with an ailment which confined her to her bed for twelve years.

Miss Price was the author of several articles and three small books. She made a collection of more than 2,000 specimens of plants, some of which she exhibited at the World's Columbian Exposition in Chicago in 1893. She also did more than 500 water-color and pencil sketches which often accompanied her specimens.

Sadie apparently maintained her dignity when collecting in the field in her long dress and high shoes. In one of her writings in *The American Botanist*, she recalled her search for the filmy fern (*Trichomanes boschianum*):

> Far under the cliff, where the rocks were dripping with moisture and where the sunshine never reached the fronds, was the home of this delicate fern. My first view of this interesting plant was a memorable one. A turn in the cliff, a lowering of the head, still lower, down on the knees, then I obtained a full view of the dainty beauty. But to collect it a humbling of my pride was necessary, as I had to cast aside hat and botanical equipment, and crawl under the projecting rock, with scarcely room for head and shoulders to enter. It meant sore muscles and a fresh accumulation of mud on the dress that had already passed recognition, but it also meant a treasure to gloat over!

Price's groundnut was discovered by Sadie Price in 1895 in rocky woods near Bowling Green, Kentucky. She collected it in flower in June and July and in fruit during August. The plant is a member of the pea family, and is one of two species in the United States which belong to the genus *Apios*. The other species, *Apios americana*, is a rather common, maroon-flowered plant of the eastern part of the country.

Apios priceana is a vine which scrambles over other vegetation with its rather slender, twining stems. Two or three elongated clusters of flowers usually arise on slender stalks from the axils of some of the leaves. Each sweet-pea-shaped flower is rose-pink in color. The largest of the petals is unique in the pea family by having a conspicuous swollen, spongy tip. Price's groundnut grows from a large, starchy, underground tuber said to get as large as six inches in diameter.

In addition to the Bowling Green collections, Price's groundnut has been found recently in Kentucky in a few far western counties by Raymond Athey, an extraordinary amateur plant sleuth from Paducah.

Athey has found Price's groundnut in two distinct habitats, and he consented to show these to me. We left Paducah for Smith-

land, where there is still great barge traffic at the mouth of the
Cumberland River, and then proceeded northeast, finally paral-
leling the Ohio River for a number of miles. Our road swerved up
to the base of some low limestone bluffs, and Athey gave the word
that we had arrived. The bluff was about thirty feet back from
the road, and the woods between the road and the bluff face had
many boulders of all sizes scattered throughout. Kentucky coffee
trees (*Gymnocladus dioica*), sweet gums (*Liquidambar styraci-
flua*), and shagbark hickories (*Carya ovata*) dominated the canopy. The understory was a scrambling entanglement of vines, giving the impression that some of the trees had been cut out in not
too distant times. When trees are cut out of a forest to open up
the canopy, viny species often "rush" in to take advantage of the
extra light that reaches the forest floor. Among the vines present
was Price's groundnut. It was July 22 and almost the perfect time
to see this species, for it was in full bloom. The conical-shaped
spikes of rose-pink flowers were gorgeous in contrast to the dark
green leaves behind them. Athey told me that in his experience
with this species, as the trees in the forest mature and cast more
shade, Price's groundnut will become less vigorous.

This was evident when we got to the second place to observe
Price's groundnut. This area was a dry woods with a gravel road
penetrating through it. At one place, a power line cut across the
woods and paralleled our road. Athey said that Price's groundnut
would be below the power line, and that following periodic clearing of the small trees from beneath the lines, the groundnut would
grow and spread more vigorously. Sure enough, Price's groundnut
was here in considerable quantity, growing over brushy vegetation
and even encroaching into the adjacent woods.

Several sightings of this species were made in rocky woods
around Nashville, Tennessee, in the 1930s. The plant today occurs
in sizable numbers in one county in eastern Tennessee.

Apios priceana has been found also in Alabama, Mississippi, and
Illinois. At its only Illinois location in the Shawnee National
Forest in Union County, it is found in a thicket at the edge of a
spring-fed swamp. The locality is in one of the most fascinating
natural areas in the country, known as the LaRue-Pine Hills,
where massive limestone cliffs tower above an extensive spring-fed
swamp. Over 1,150 different kinds of flowering plants occur here
in little more than eight square miles. As a result of this botanical

diversity, LaRue-Pine Hills in 1973 was designated the first Ecological Area in any of our National Forests.

Because of the limited number of locations for this species, it should be considered for federal listing.

Florida Tree Cacti or *Cereus robinii* var. *robinii* and *Cereus robinii* var. *keyense.*

Cactus family. Tall succulents to 18 feet; stems with 9–10 vertical ribs; spines stiff, yellow, ½ inch long, in clusters of 10–14 or more. Flowers light green, to 3 inches long; sepals and petals several. Season: April.

Not all of the tall cacti of the United States grow west of the Mississippi River. There are two remarkable tree cacti that live in the Florida Keys. One of these occurs in the lower Keys, where it was found as early as the 1830s. It was originally called *Cephalocereus keyense.* The other was discovered on Lower Matecumbe Key by John Soar and Charles T. Simpson about 1907 and ultimately named *Cephalocereus deeringii,* or Deering's tree cactus, in 1917 by John Kunkel Small, the great authority on the flora of the southeastern United States.

Subsequent critical study of these two tree cacti has revealed that they are varieties of the same species, whose binomial has been changed to *Cereus robinii.* The Lower Matecumbe Key plant, or Deering's tree cactus, became known as var. *robinii,* and the lower Keys plant was called var. *keyense.*

Cereus robinii var. *keyense* was threatened with extinction by the beginning of the twentieth century. In 1917, Dr. Small reported it "on the verge of extermination in its natural habitat." Small continued, in the *Journal of the New York Botanical Garden,* that

in recent years, with the destruction of the [Key West] hammock for securing firewood and for developing building sites, this interesting cactus has become scarce. A dozen years ago, many fine and robust trees, some of them over twenty feet high, stood in the hammock east of the city. These have long since disappeared. The writer has counted a score of prostrate skeleton remains scattered over the hammock floor which were either cut down by wood hunters or blown down by storms. Many small plants and a few small trees remained standing as late as 1915. The trees have now been destroyed, and nearly all of the small plants have been exterminated during the past year by the construction of a baseball field and by the clearing of building lots.

For many years, the lower Keys tree cactus was considered extinct. Then, in 1973, Brian Lamb, a noted cactus authority from Great Britain, along with colleague Pete Forrester, reported that they had rediscovered this remarkable tree cactus on a densely wooded key, growing in wet leafmold less than fifty yards from a swamp. The habitat is most unpleasant to walk in because of a dense entanglement of thorny plants and innumerable meshes of spider webs. In addition, large diamondback rattlesnakes patrol the area. The largest plants of the lower Keys tree cactus that Lamb reported were about eighteen feet tall.

The lower Keys tree cactus is apparently very rare in the wild and must be given complete protection from collectors and urban expansion of the Keys.

Deering's tree cactus (*Cephalocereus robinii* var. *robinii*) from the Lower Matecumbe Key area has apparently had even more difficult times from the threat of collectors, hurricanes, building contractors, and wood hunters.

Dr. Small reported that Deering's tree cactus was common in 1916 on the coral rocks exposed on Lower Matecumbe Key, but that all the tall specimens were prostrate on the hammock floor as a result of hurricanes. Intensive searches for this cactus have failed to find it on Lower Matecumbe, but Brian Lamb reported finding it on an adjacent key.

Deering's tree cactus was a plant to behold! Its slender, erect stems sometimes grew to a height of thirty feet, but the few known specimens from the wild today are little more than three feet tall. The specimens that exist today apparently have not flowered for several years, but Dr. Small reported in 1917 that this plant had three-inch-long light green flowers that bloomed during the afternoon. He also described the fruits as dark red and fleshy and nearly two inches in diameter, containing several shiny seeds.

Charles Deering, for whom this tree cactus gets its common name, was one of the first persons to promote the preservation of the natural features of the Florida Keys. He planted a number of these tree cacti in the Miami area.

Because the two Florida Keys tree cacti are apparently nearly extinct in the wild, they should be given a place on the list of federally endangered plants.

Oconee Bells or *Shortia galacifolia.*

Diapensia family. Perennial. Leaves basal, nearly round, toothed, to 3 inches. Flower solitary on a leafless stalk, pinkish or whitish; sepals 5, green; petals 5, to 1 inch. Season: March, April. (See color plate 11.)

I was completing several days' study in South Carolina's Sumter National Forest in the Blue Ridge Mountains and was still tingling with excitement at all the remarkable events I had experienced.

I had explored along the wild and awesome Chattooga River, which separates South Carolina from Georgia. The Chattooga, now a designated Wild and Scenic River and the setting for the motion picture *Deliverance,* flows with varying intensity from Russell Bridge to Earl's Ford. Canoeists rate three levels of difficulty on the Chattooga. Class three, the most severe, has numerous high, irregular waves and rapids with passages that are clear, though narrow, requiring expertise to maneuver. That last phrase convinced me to confine my exploration of the Chattooga to the riverbanks.

I had hiked through the cathedral woods of hemlock and pines to Ellicott Rock, the historical boulder that marks the point where South Carolina, North Carolina, and Georgia come together.

I had stared in amazement in the Long Cane Scenic Area southeast of Abbeville at the sixteen-foot-tall colony of giant cane (*Arundinaria gigantea*), one of the few native woody grasses of the United States.

I still had one goal left before starting back to Illinois. It was to see the beautiful but rare Oconee bells (*Shortia galacifolia*), a member of the rather obscure Diapensiaceae family. One of the Sumter District rangers showed me two specimens growing in the front yard of a rural resident along South Carolina Highway 107, so I knew what I was to look for.

I was on my way to a gorge which had been carved by a now placid, picturesque mountain stream. The area had been settled nearly two centuries ago, but most of the landowners had moved out by 1900. Evidence of these early settlers could be seen in the fallen chimneys, wagon roads, and an old discarded tool here and there. A small row of rough, hand-hewn tombstones, some with no indication of who was buried below, was seen.

Along the stream, in an area heavily shaded by rhododendron, was a colony of Oconee bells. All of the nearly round leaves are clustered at ground level. From among the cluster of basal leaves, from late March to mid-April, a slender, leafless stalk rapidly lengthens to about six inches and forms a solitary pinkish or whitish flower.

The history of *Shortia* is fascinating. The plant was first discovered in the Carolina mountains in 1787 by André Michaux. Michaux was a French botanist who spent several years in the eastern United States searching for potentially useful plants for his government. He explored as far west as the Mississippi River and found many species new to science, most of which he named in his *Flora Boreali-Americana*, which was published in 1803 after he returned to France.

Michaux failed to identify the Oconee bells, however, because when the American botanist Asa Gray went to Paris in 1839 to study the Michaux collections, he found the specimen of Oconee bells lying unnamed in the herbarium. Realizing it to be new to science, Gray called the plant *Shortia galacifolia*. *Shortia* was to commemorate Gray's friend and colleague from Kentucky, Charles W. Short, and *galacifolia* indicated that the plant had leaves that resembled those of *Galax*, a common mountain herb of the Appalachians.

On his return to the United States, Gray attempted to relocate *Shortia galacifolia* in the wild. With only Michaux's note "high mountains of Carolina" to guide him, Gray unsuccessfully explored Grandfather Mountain, Mount Mitchell, and other mountains in the vicinity for this plant.

Finally, nearly one hundred years after its original discovery, seventeen-year-old George Hyams of Statesville, North Carolina, stumbled across the Oconee bells on the banks of the Catawba River in 1877.

It seems inconceivable today that the rediscovery of *Shortia* could create the excitement that it did. Naturalists and non-naturalists alike rushed to the area to get specimens. A good, well-pressed specimen sold for as much as fifty dollars. It was no wonder that the plants at the new locality became extremely scarce.

In the meantime, exploration of additional wild gorges, particularly in Oconee County, South Carolina, has revealed the presence of several other localities for this plant, although some of these sites have been destroyed by large impoundments which were created several years ago. More recently, a few plants have been found in Rabun County, Georgia.

Although Oconee bells is not threatened at the moment with extinction, care should be taken that the plants that remain are not overcollected and that no more of the gorges where it occurs are inundated.

Ruth's Golden Aster or *Heterotheca ruthii.*

Aster family. Perennial; stems much branched, to 10 inches. Leaves very narrow, without teeth, silvery-hairy, to 2 inches. Flowers crowded together in glandular heads, with bright yellow rays ⅓ inch long, surrounding a small yellow disk. Season: August to October.

The Hiwassee River originates in the mountains of North Carolina and northern Georgia and flows westward through the Unicoi Mountains of southeastern Tennessee. It is a beautiful, fast-flowing, clear stream typical of the Appalachian Mountain range. In recent years, the flow of the water has been altered by the development of large impoundments such as Hiwassee Lake and Appalachia Lake. Along much of its course, however, the river has retained its pristine character. Several miles after the Hiwassee River enters Tennessee, it has carved a scenic gorge where rugged cliffs line the river. The Hiwassee Gorge is in the Cherokee National Forest, a massive region of wild and scenic beauty. The Cherokee occurs for nearly 200 miles on the Tennessee side of the Tennessee–North Carolina border, interrupted by a 50-mile segment of the Great Smoky Mountains National Park. The plant life in the forest is rich in the number of species present.

Prof. Albert Ruth of the University of Tennessee was studying this mountain vegetation in 1894 when he discovered a golden-flowered aster in the rocks along the Hiwassee Gorge. Although similar to another golden aster (*Heterotheca graminifolia*) that was growing nearby, Ruth's plant was obviously different. Recognizing the distinctness of the new plant, Dr. John Kunkel Small of the New York Botanical Garden described it as a new species in 1897, naming it after Professor Ruth. It is known as *Heterotheca ruthii,* or Ruth's golden aster.

Professor Ruth returned to the Hiwassee Gorge many times, for there are several collections of Ruth's golden aster made by him between 1894 and 1902. Ruth was apparently the only person ever to have collected the plant, and in 1969 one botanist who was studying the golden asters suggested that Ruth's golden aster was probably extinct.

At about the same time, Frank Bowers of the University of Tennessee, who had unsuccessfully explored the Hiwassee River for *Heterotheca ruthii,* rediscovered Ruth's golden aster growing in crevices of rocks above the Hiwassee River. For nearly a mile and a half along the river, Bowers found plants of this species. They were all growing from a rocky exposure known as phyllite. Much of the phyllite was covered by an assemblage of rock-inhabiting lichens and mosses. Very few flowering plants were in the immediate vicinity of the golden aster.

In 1976, Andrew J. White of the University of Tennessee was able to extend the range of Ruth's golden aster to three miles along the Hiwassee River. He also found a second location for this species a few miles south along the Ocoee River. A search of other likely rivers in the area failed to turn up additional localities for the golden aster, however.

Heterotheca ruthii is a much branched, low-growing perennial with slender stolons at the base. The golden flowers are crowded into inch-wide heads with a dozen or so bright yellow rays. The rays, which usually have a slight notch at their tip, are about one-third inch long.

The few known colonies of Ruth's golden aster are under constant threat from construction activities, chemical pollution, and overcollecting. This species is currently under review by the United States Fish and Wildlife Service for possible listing as a federally endangered or threatened species.

Cumberland Rosemary or *Conradina verticillata.*
Mint family. Slender shrub to 1 foot; stems square. Leaves simple, whorled on the stem, dotted, to 1 inch. Flowers few,

*lavender or purple, less than 1 inch long; petals 2-lipped; stamens
4. Season: May, June.*

During March 1979, Rob Farmer, president of the Tennessee Native Plant Society, announced that this organization would embark
on a project to find and map all wild populations of a low-growing
shrub of the mint family known as *Conradina verticillata*, the
Cumberland rosemary.

This species is endemic to the Cumberland Plateau of north-
central Tennessee and adjacent Kentucky, and was known from .
very few localities.

The original collection of this mint was made by Prof. Albert
Ruth of the University of Tennessee in 1894, the same year,
coincidentally, that he discovered Ruth's golden aster (*Heterotheca ruthii*). At the time that Ruth found the mint, he identified
it as a *Conradina*. There was only one *Conradina* known from
the southeastern United States, *Conradina canescens* of the
Coastal Plain. Botanists thought that Ruth's collection represented a disjunct population of *C. canescens*.

In 1930, Dr. H. M. Jennison of the University of Tennessee
was taken to Morgan County, Tennessee, by a Mrs. Charles
Brooks to look at a plant Mrs. Brooks had found. It turned out
to be another specimen of Ruth's *Conradina*. Jennison became
puzzled by the fact that this *Conradina* was a mountain plant,
while all the other known collections of *Conradina canescens*
were from the flat Coastal Plain.

The following spring, Jennison and Dr. A. J. Sharp returned to
Morgan County to observe and collect the *Conradina*. Jennison
became convinced that it was, indeed, different from the Coastal
Plain species, and he named it as a new species, *Conradina verticillata*, in 1933.

In 1934, E. Lucy Braun, one of the foremost ecologists of the
Middle West, who was gathering data for her monumental *Deciduous Forests of the Eastern United States*, discovered the Cumberland rosemary along the South Fork of the Cumberland River
in Kentucky. Later, the plant was found along Caney Fork and
the New River, all in north-central Tennessee. Until 1965, all of
the known locations for this species were in the Cumberland River
drainage.

Dr. A. J. Sharp and others in 1965 discovered *Conradina verti-*

cillata in the Tennessee River drainage system, and by 1979 it was known from four different rivers and creeks in that system.

The Tennessee Native Plant Society, led by Dennis D. Horn, who catalogued and plotted every known site of *Conradina verticillata*, documented this species from five Tennessee counties and one Kentucky County.

This species is called Cumberland rosemary because the leaves, when crushed, have a scent similar to that of the European rosemary (*Rosmarinus officinalis*), a commonly grown culinary herb.

This plant has many branches which spread from the center of the plant before becoming ascending. On close observation, the inch-long leaves may be seen to possess small dots, indicating the resinous glands which give the leaves their characteristic aroma. From among the uppermost leaves are borne one to seven lavender or purple flowers which bloom in late May and early June.

The Cumberland rosemary is listed as endangered by the Committee for Tennessee Rare Plants, and is under consideration by Kentucky authorities. It is being reviewed for federal listing.

Appalachian Avens or *Geum radiatum*.

Rose family. Perennial. Most leaves basal, nearly round, with low teeth. Flowers several at tip of the 18-inch, nearly leafless stalk, bright yellow; sepals 5, green; petals 5. Season: June to August.

For many years after I became interested in plants, it was my desire to view the Catawba rhododendron in its fullest purple splendor on Roan Mountain astraddle the North Carolina–Tennessee state line. Dr. Elisha Mitchell, for whom nearby Mount Mitchell was named, said that "Roan is the most beautiful of the high mountains—with Carolina at the viewer's feet on one side and Tennessee on the other, and a green ocean of mountains rising in the tremendous billows immediately around him."

I had been to Roan Mountain twice in conjunction with research I was doing in the Cherokee and Pisgah national forests. The first time, the Catawba rhododendrons were just finishing their early summer display. A few blossoms were still hanging on, tempting me to come back another year. The next time, the flower buds were beginning to swell, for I was nearly a week early.

During the spring of 1972 I began to lay plans for a foray to Roan Mountain, specifically to see the Catawba rhododendron.

Since 1947 the citizens of the community of Roan Mountain, Tennessee, have held an annual rhododendron festival. I figured that if anyone could give me a close estimate for the blooming time of the rhododendron, it would be the people who were charged with scheduling the festival. Knowing no one in Roan Mountain, I telephoned the Roan Mountain post office, and the postmaster told me the festival was to be held June 16 and 17, although he couldn't guarantee one hundred percent that the rhododendrons would be at their peak.

I took my chances. My family and I pulled into town late in the afternoon of the fifteenth. Everyone was buzzing with excitement. So was I after I noticed that several of the festival decorations were using fresh Catawba rhododendron flowers. Each table in the café where we ate was adorned with a water glass holding a cluster of the flowers. It meant that they were in bloom up on the mountain!

After a hurried breakfast the next morning, we took Tennessee Highway 143 and twisted our way to the summit. A short side road brought us to the parking lot beyond Roan High Knob. There before us were the rhododendron gardens, touted as the most splendid natural display of vegetation in the world. I agreed. Never had I seen such a breathtaking sight! Rounded rose-purple mounds of Catawba rhododendron in a parklike setting with a background of pointed-crowned balsam fir and spruce greeted our eyes. Far in the distance, in the blue haze of the morning, were 110 peaks of the Blue Ridge Mountains. (At least that's how many one of the local residents told us we could see from the summit.)

Explorers as early as the 1780s had written about the beauty of Roan Mountain. In 1877, former Union Army general Thomas Wilder built a 20-room log building on the summit, where he entertained friends and visitors. "Goin' to the Roan" became so popular that Wilder replaced the log structure in 1855 with a 166-room, three-and-a-half story hotel he called Cloudland. The dining room of the hotel straddled the state line. Stagecoaches and carriages brought guests to Cloudland by way of a narrow, steep trail. When the resort hotel ceased to operate around the time

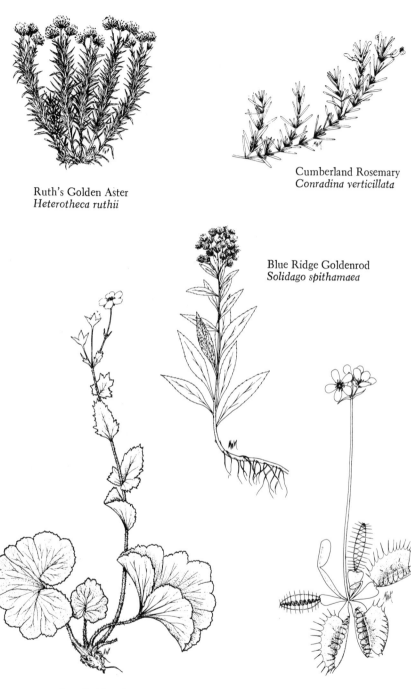

Ruth's Golden Aster
Heterotheca ruthii

Cumberland Rosemary
Conradina verticillata

Blue Ridge Goldenrod
Solidago spithamaea

Appalachian Avens
Geum radiatum

Venus Fly Trap
Dionaea muscipula

of the First World War, the mature spruces and firs on the summit were harvested, and many of the rhododendrons were sold as nursery stock. After Roan Mountain was acquired by the United States Forest Service in 1940, the rhododendrons made a comeback and have developed into the beautiful display on Roan Mountain today.

One of the visitors to the summit of Roan Mountain during the latter part of the nineteenth century was Nathaniel Lord Britton, founder of the New York Botanical Garden. While on the mountain, Britton found a new species of false goatsbeard (*Astilbe crenatiloba*), which has not been seen since.

Several other rare plants occur on Roan Mountain and adjacent high mountains of North Carolina and Tennessee. One of these is an attractive yellow-flowered herb in the rose family known as the Appalachian avens (*Geum radiatum*).

This species occurs in rocky openings at higher elevations in five counties in North Carolina and three in Tennessee. It is a perennial with a cluster of basal leaves which resembles leaves of the florist's geranium. Each stem is terminated by a small cluster of yellow flowers which open during the summer.

Although the Appalachian avens is rather common in a few localities, its very limited habitat makes it susceptible to disturbance and ultimate extinction. It is listed as endangered in both North Carolina and Tennessee.

Blue Ridge Goldenrod or *Solidago spithamaea.*

Aster family. Perennial; stems to 1 foot, hairy. Leaves alternate, simple, elliptic, coarsely toothed, to 2 inches. Flowers crowded together in ½-inch wide heads; rays yellow, very short. Season: July, August.

A plant even more uncommon than the Appalachian avens and growing on the open treeless summits of some of the Blue Ridge Mountains in the vicinity of the avens and the rhododendron, is a summer-flowering goldenrod known as the Blue Ridge goldenrod, *Solidago spithamaea.*

These treeless areas on the high slopes and summits of several of the southern Appalachian mountains are referred to as balds. Some of these balds support primarily grasses and herbs. Others, which are dominated by rhododendrons, azaleas, and mountain laurel, are called heath balds. Because these mountain summits are not at a high enough elevation to be above tree line, the absence of trees has caused considerable speculation among naturalists.

Since the early settlers reported the presence of these balds, logging as a cause for treelessness can be ruled out. One possible explanation, however, is the presence of wildfires which may have suppressed most trees while stimulating rhododendrons and mountain laurels to resprout quickly. Whatever the reason, the balds are unique in the flora they support.

The Blue Ridge goldenrod is one of these unique species on balds and in rocky crevices in the mountains of North Carolina; it is known from only three counties. There is a report of this species from Carter County, Tennessee, but authorities in that state believe the Blue Ridge goldenrod no longer occurs there.

Solidago spithamaea is an extremely dwarfed goldenrod rarely a foot tall. Several heads of yellow flowers are arranged in a narrow, crowded cluster that terminates the stem.

A federal listing seems appropriate for the Blue Ridge goldenrod because of its extreme rarity.

Venus Fly Trap or *Dionaea muscipula.*

Venus fly trap family. Perennial. Leaves insectivorous, hinged, with spiny teeth along the edges. Flowers several on a 6- to 10-inch leafless stalk; petals 5, white; stamens about 15. Season: May, June.

Venus fly trap, with its insect-entrapping mechanism, evokes more interest than almost any other kind of plant. Its unusual appearance and the miraculous way it reacts to stimuli make it difficult to realize sometimes that this is a plant that grows in the wild. And the wild places where it occurs are primarily in the Coastal Plain of North and South Carolina, where it should be allowed to remain.

The novelty of the Venus fly trap has caused it to become greatly reduced in numbers for commercial purposes. Although it has been found in seventeen counties in North Carolina, three counties in South Carolina, and very recently in one county in the Florida panhandle, it is rare in several of them. In the wild, the

Venus fly trap lives in open bogs, savannas, and in wet sandy ditches.

The entrapment process is ingenious. Each trap is actually a modified leaf folded lengthwise. It has a winged leaf stalk which abruptly tapers to a short projection before joining the trap. Each half of the trap has inward-pointing, stiff, spinelike teeth along its outer edge. Just below these teeth is a row of glands which secrete nectar to attract insects.

On the inner face of each half of the trap are three or four sensitive bristles. One or more of these bristles has to be touched twice within a period of about twenty seconds before the trap is activated to close. This is thought to be a "false alarm" mechanism so that if some wind-blown object happens to touch a bristle, the trap will not automatically close. This is important, because it takes about twenty-four hours for the trap to reopen, requiring an unnecessary expenditure of energy if no insect is present.

If an insect does activate the trap, the trap closes in a quick, two-stage process. The first step involves the two halves of the trap coming together only tightly enough for the teeth of the trap to interlock, but loosely enough to allow very small insects to crawl back out. This again is seen to be an efficiency measure on the part of the plant, since the amount of nutrients that the plant could obtain from a tiny insect would not be worth the energy involved in going through the complete process of closing and opening the trap. In addition, since each trap opens and closes from insect stimulation only three or four times during its lifetime, it is essential that it doesn't waste one of the few catches it is capable of on the entrapment of a small insect.

After this instantaneous shutting of the trap, the second step of the process begins. It involves a slower squeezing together of the two halves of the trap, usually crushing the insect. As soon as the trap is completely sealed, digestive enzymes are secreted from special cells on the inner faces of the trap. These enzymes begin to decompose the prey, and nutrients are absorbed by the plant. When the process is completed, the trap opens and exposes the undigested parts of the insect.

While indiscriminate collecting of the Venus fly trap for commercial purposes has decimated the population, other threats from habitat destruction and encroachment of forests need to be closely checked.

Dr. George Folkerts of Auburn University has stated that current forest practices are reducing the number of Venus fly trap colonies. Many of its former habitats are now covered by even-aged pine stands. Preparing a site for a new pine plantation completely eliminates any fly traps that are present. Dr. Folkerts goes on to say that even without site preparation, the shading caused by the pines, along with the suppression of fires, destroys fly trap populations.

While most of the interest in the Venus fly trap centers on the leaves, the plant does produce attractive white flowers that bloom during May and June.

Dionaea muscipula is listed as a threatened species in North Carolina, but it has no current federal status.

Southern Yellow Orchid or *Platanthera integra.*

Orchid family. Perennial; stem 1 to 1½ feet tall. Leaves 1–2, folded lengthwise, to 8 inches. Flowers bright yellow, to ½ inch long, crowded in a spike. Season: July to September. (See color plate 12.)

Some of the most beautiful orchids in the United States belong to a genus of plants that used to be known as *Habenaria*, but which most botanists today refer to as *Platanthera*. (The name *Habenaria* still refers to some southern orchids, however.)

The differences between the many species of *Platanthera* in North America involve one of the highly modified petals of the flower, known as the lip.

In some Platantheras, the lip is fringed along its outer edge, giving to these species the common name of fringed orchid. In others, the lip is without a fringe, although it may be toothed or very shallowly lobed. These are the fringeless orchids. *Platanthera integra*, the southern yellow orchid, belongs to this latter group. When in flower, few orchids can match the beauty of the southern yellow orchid. As many as sixty half-inch-long, bright yellow flowers crowd together in a terminal cluster.

Although this orchid occurs in a band across the southern United States from the southeastern corner of Texas to eastern North Carolina, the original collection was made from disjunct populations found in New Jersey. Thomas Nuttall described this orchid in 1818, based on plants from what he called a swampy habitat in New Jersey. It actually occurs in wet barrens and sphagnum bogs. Despite diligent search by New Jersey naturalists, only a very few sites are currently known for the southern yellow orchid from that state. It is considered endangered in New Jersey.

From New Jersey, the southern yellow orchid seems to jump to eastern North Carolina, where it grows in swamps and flatwoods of the Coastal Plain, as well as in a few southern mountains in the state. It frequently occurs where pitcher plants live. It is a species of special concern to North Carolina botanists because of the elimination of much of its habitat.

Platanthera integra is also uncommon to rare in the South Carolina Coastal Plain, as well as in Tennessee, Georgia, northern Florida, Alabama, southern Mississippi, southern Louisiana, and southeastern Texas.

Even though the southern yellow orchid is found in several states, the fact that it is rare throughout its entire range has brought it under review by the United States Fish and Wildlife Service.

Leafy Purple Prairie Clover or *Petalostemum foliosum.*

Pea family. Perennial; stems to 3 feet. Leaves alternate, compound, divided into 11–31 narrowly oblong leaflets. Flowers purple, in 1-inch-long spikes; petals sweet-pea-shaped. Season: July, August. (See color plate 13.)

We had just picked up Dr. Paul Somers at the Tennessee Heritage Program office in Nashville for a trip out to the cedar glades east of the city to see the endangered Tennessee coneflower (*Echinacea tennesseensis*).

On our way out of town, Paul asked if I would like to see *Petalostemum foliosum,* the leafy purple prairie clover. Would I ever! I had been interested in this rare member of the legume family for a number of years. It is a species that was first found more than a century ago about fifty miles west of Chicago along the banks of the Fox River. Sometime later, Rev. E. J. Hill found it on Altorf Island in the Kankakee River of northeastern Illinois. Altorf Island is the same place where Hill also discovered the very rare Kankakee mallow (*Iliamna remota*). During the next several years after its original discovery, the leafy purple prairie clover was found in a half dozen other areas on gravelly soil in the northern fourth of Illinois. By 1970, however, Illinois botanists were unable to locate a single plant of this species anywhere in the state. But on August 17, 1974, Gerould Wilhelm of the Morton Arboretum at Lisle, Illinois, rediscovered *Petalostemum foliosum* in Illinois. It was growing in a prairie habitat along with such prairie species as little bluestem (*Schizachyrium scoparium*), Indian grass (*Sorghastrum nutans*), wild bergamot (*Monarda fistulosa*), mountain mint (*Pycnanthemum virginianum*), and the stiff prairie goldenrod (*Solidago rigida*).

In the meantime, and surprisingly, this same species of prairie clover had been found on a few limestone glades in a few counties in Tennessee surrounding Nashville. This habitat is unlike the Illinois locations. The cedar glades of Tennessee have pieces of white limestone scattered across the flat terrain. Most of the plants that grow with the leafy purple prairie clover on the Tennessee glades are not found in northern Illinois. There is, however, one other plant which grows in the limestone glades in Tennessee and also, at least until a few years ago, in the prairies of northern Illinois where Reverend Hill collected. It is the yellow-flowered Tennessee milk vetch (*Astragalus tennesseensis*). The unusual parallel distribution of these two species in northern Illinois prairies and central Tennessee glades is remarkable.

We came to a major intersection in Nashville and pulled into a nearby parking lot. One of the quadrants at the intersection was a vacant lot, uninvitingly grown up with brush. As we pushed our way through the vegetation (which we later found out was covered with chiggers), we saw white gravelly rocks showing through the dense plant growth underfoot, indicating we were in a badly disturbed limestone glade. Despite the encroachment of weedy vegetation, there was the leafy purple prairie clover, some five or six plants in all. They were in full bloom with their attractive short spikes of purple flowers.

Petalostemum foliosum stands two or three feet tall. Several purple flowers which bloom during July and August are crowded into inch-long conical spikes.

In addition to the single Illinois location and the sites in four counties in central Tennessee, the leafy purple prairie clover has been found recently in a wet limestone glade in northern Alabama. This species is so rare that the United States Fish and Wildlife Service is reviewing it for possible listing as a federally endangered species.

4

Diminishing Wildflowers
of the North-Central States

The north-central United States, as a group, has fewer rare species than any other area in the country. Part of this can be attributed to the absence of rugged mountain ranges and other unusual habitats where rare plants often lurk. The northern monkshood, which is one of two plants in the north-central states that is on the federal list, also occurs in New York.

The states considered to be in the north-central region are Ohio, Indiana, Michigan, Wisconsin, Minnesota, Illinois, Iowa, Missouri, North Dakota, South Dakota, Nebraska, and Kansas.

Northern Monkshood or *Aconitum noveboracense*.
Buttercup family. Perennial; stems to 3 feet. Leaves alternate, deeply lobed. Flowers purple, to 1 inch; sepals 5, showy, one of them hood-shaped; petals 2, small, concealed by the hooded sepal. Season: June, July. (See color plate 14.)

On April 26, 1978, the northern monkshood (*Aconitum noveboracense*), one of the rarest flowering plants in the United States, was designated a federally threatened species. This action became necessary when the very attractive perennial herb was reduced to only fourteen living colonies of plants, and some of these were being threatened by a proposed impoundment that would inundate their habitat.

One of the interesting features about the northern monkshood is the disjunct distribution of these colonies. Most of them are known from northeastern Iowa and adjacent southwestern Wisconsin, but there are two populations in northeastern Ohio and three in New York's Catskill Mountains.

Aconitum noveboracense was described by Asa Gray in 1886 from specimens collected prior to 1860 from a site in Chenango County, New York. Despite repeated efforts to relocate this species in Chenango County, it has not been found there since about 1890.

However, around 1870, the northern monkshood was discovered in the Catskill Mountains of New York, less than one hundred miles from the original Chenango County site. Today, this species occurs in three localities in the Catskills, where it grows in steep and wooded streamside habitats, on shaded cliffs, and in semishaded seepage areas.

The first report of the northern monkshood from Ohio was made in 1870 in the vicinity of Akron, where it still occurs. A second station in Ohio exists in Portage County. Here the plants are found in a deep gorge, where they grow in crevices on the lower portions of sandstone cliffs.

Although four great glaciers descended down across northern North America during the Ice Age, each of them bypassed a small area of northeastern Iowa, southeastern Minnesota, northwestern Illinois, and southwestern Wisconsin. This region has become known as the "Driftless Area," and several kinds of plants have taken refuge in this area which do not occur in the adjacent glaciated land. Several of these Driftless Area plants are also found in the arctic regions of North America or on western mountains. These include the moschatel (*Adoxa moschatellina*), Lapland rosebay (*Rhododendron lapponicum*), and arctic primrose (*Primula mistassinica*).

The northern monkshood is one of those species found in the Driftless Area. At its seven known locations in Iowa and five in Wisconsin (a sixth Wisconsin location no longer supports this species), it lives in a cold soil environment at the bases of cliffs or on talus slopes. There is usually a continuous drainage of cold air from the nearby bedrock. Considerable flowage of cold groundwater keeps the habitat moist.

The first Wisconsin report for this species was just prior to 1900, while the initial collection in Iowa was not made until shortly after the turn of the century.

Several threats to its survival keep the northern monkshood in hot water (a predicament for a cold-water species!). Some localities are adjacent to well-traveled trails; one is in a city park (albeit rather inaccessible); one grows next to a state highway; and one is in a power-line corridor. There is also danger that logging activities will open up the forest canopy, permitting too much sunlight to reach this shade-loving species. Because the plants have attractive flowers, they are subjected to transplanting from the wild into gardens.

In 1970, the United States Army Corps of Engineers proposed a reservoir for flood control and recreation purposes along the Kickapoo River in southwestern Wisconsin, where several colonies of the northern monkshood exist. The corps sponsored a survey to determine the importance of this plant in the area. The survey provided the Fish and Wildlife Service with data to determine that *Aconitum noveboracense* should be given threatened status on the federal list of endangered and threatened species.

Aconitum, a genus in the buttercup family, is comprised of about seventy species found throughout the north temperate regions of the world. All of these plants have a unique flower which is helmet-shaped with a caplike hood at the top. Several members of the genus produce various kinds of alkaloids which have important medicinal properties. It is not known whether the northern monkshood has these medicinal alkaloids present, but in all likelihood it does.

The northern monkshood is a perennial which may reach a height of two to three feet. The attractive inch-long purple or blue flowers bloom in June and July.

Lake Iris or *Iris lacustris*.

Iris family. Perennial. Leaves basal, elongated, to ½ inch broad. Flowers blue to purple; sepals 3; petals 3; stamens 3. Season: May to July. (See color plate 15.)

Two years after British-born Thomas Nuttall came to America in 1808, he entered into the employ of Prof. Benjamin Smith Barton, who was seeking a competent young botanist to search the "wilds of the Northwest" for plants which Barton could include in a flora of North America.

Promised eight dollars per month plus expenses, Nuttall headed west, traveling on foot and by boat past Lake Erie and Lake Huron to Detroit, exploring around this infant city for nearly a month. On July 29, 1810, he left Detroit for Mackinac Island by canoe, and then on to Green Bay, Wisconsin, where he arrived in late August.

Nuttall's pioneering exploration along the Great Lakes turned up more than a score of plants new to science. Among those new species discovered in the vicinity of Michilimackinac, Michigan, was the Lake Huron tansy (*Tanacetum huronense*) and the lake iris (*Iris lacustris*).

The lake iris deserves special attention. It occurs almost exclusively along the northern shores of Lake Michigan and Lake Huron, where it grows in moist sands and gravels and in limestone crevices, although there is one station for it in Door County, Wisconsin, and a few colonies on Manitoulin Island and the Bruce Peninsula in Ontario.

After Nuttall's original collection of *Iris lacustris* in 1810, the species was found again by Capt. David Bates Douglass on the Cass Expedition of 1820, on which Douglass served as zoologist, botanist, and cartographer. The Cass Expedition was the first organized scientific expedition into Michigan, conceived and organized by territorial governor Lewis Cass.

The lake iris is known in Michigan today from fewer than a dozen counties. All but a few of the localities are along the shore of either Lake Michigan or Lake Huron. There are two inland populations—one near the Menominee River in Menominee County on a sandy gravel ridge, and one on the calcareous banks of the Escanaba River in Delta County.

On the Door Peninsula of Wisconsin, *Iris lacustris* is found on sandy beach ridges and under conifers in thin humus, where it sometimes grows in large colonies. A previously known locality in Milwaukee County has apparently been destroyed.

This iris is similar in appearance to the dwarf crested iris (*Iris cristata*), a plant native to the southern United States. The lake iris differs by its usually narrower leaves and sepals, its shorter floral tube, and a slightly longer seed-containing capsule. The seeds are dispersed by ants.

Iris lacustris is a perennial which has very slender underground rhizomes. The blue to purple flowers bloom during May and early June. The half-inch-long capsules are three-angled.

Lake iris has threatened status in Michigan and endangered status in Wisconsin. It is being reviewed by the United States Fish and Wildlife Service for possible federal listing.

Great Lakes Thistle or *Cirsium pitcheri.*
Aster family. Perennial; stems to 3 feet, covered with white felt. Leaves alternate, simple, deeply lobed, weakly spine-tipped on each lobe. Flowers crowded together in 1½-inch-high heads, cream-colored, the head surrounded by spine-tipped bracts. Season: May to September. (See color plate 16.)

A most distinctive and handsome thistle grows on stabilized sandy shores and on sand dunes along Lake Michigan and Lake Huron, with a single location along the shore of Lake Superior. Known as the Great Lakes thistle (*Cirsium pitcheri*), this species occurs in Michigan, Indiana, Wisconsin, and southern Ontario. It grew at one time in Illinois on sand dunes near Lake Michigan, but it disappeared from that state about 1908.

The plant commemorates its discoverer, Dr. Zina Pitcher, who first found it in 1826 or 1827, "on the great sand banks of Lake Superior." Dr. Edward Voss of the University of Michigan reveals that the great sand banks could only be the Grand Sable Dunes. It is ironic that the first collection of this species is from the only locality known for it along Lake Superior.

Zina Pitcher was a physician, obtaining his medical degree from Vermont's Middlebury College in 1822. Pitcher became an army surgeon and was assigned to the Michigan Territory, where he served at Fort Saginaw, Fort Brady (now Sault Ste. Marie), and Fort Gratiot. It was apparently during his stay at Fort Brady that he collected the thistle that bears his name.

For eight years of the nearly fifteen years he served as an army surgeon, Dr. Pitcher was stationed in Michigan. When he resigned from the service in 1836, he returned to his beloved Michigan, practicing medicine in Detroit. As a leading citizen of the state, Pitcher assisted in founding the Historical Society of Michigan, was a member of the first board of regents of the University of Michigan and founded that university's medical school, in which he taught until 1872, and served as mayor of Detroit.

Shortly after Pitcher's discovery of the Great Lakes thistle, a second collection of this plant was made during the summer of 1827 from Mackinac Island, apparently by Dr. Edwin James, also a graduate of Middlebury College and also an army surgeon stationed at Fort Brady.

Cirsium pitcheri still occurs in fairly good numbers in Michigan, where it has been found in about twenty counties, and in north-western Indiana, where it occurs along the shores of Lake Michigan. In Wisconsin it is found on sand dunes and stabilized beaches along Lake Michigan in Sheboygan, Manitowoc, and Door counties. It is listed as threatened in all of these states. The United States Fish and Wildlife Service is reviewing it for possible federal listing.

The Great Lakes thistle stands two to three feet tall. The flower heads are about one and one-half inches high and bear cream-colored flowers during the summer.

Houghton's Goldenrod or *Solidago houghtonii.*

Aster family. Perennial; stems to 2 feet, nearly smooth. Leaves alternate, simple, elongated, to ½ inch broad. Flowers crowded together into several small heads; rays yellow. Season: August. (See color plate 17.)

I have often thought that the group of summer-flowering herbs that is best identified with the flora of the eastern United States is the goldenrod. From late June through October, from the Mississippi River to the Atlantic Ocean, goldenrods raise their small heads of yellow flowers for all to see.

There are about one hundred species of goldenrods in the world. Most of them are North American, and more than seventy-five grow in the eastern states. Goldenrods may be found in woods, on bluffs, in prairies, as an integral part of old fields, and in wetlands. Although several kinds of goldenrods grow along roadsides with a number of coarse European weeds, they are all native members of our flora.

Many goldenrods are notorious for their difficulty in being identified; some of them are poorly defined, others are promiscuous and hybridize with other species.

Houghton's goldenrod (*Solidago houghtonii*) of the Great Lakes region fails to follow this pattern. It is a well-defined species, living in a well-defined habitat. The goldenrod is named for one of its discoverers, Douglass Houghton.

Douglass Houghton came to Detroit from New York in 1830 when he was twenty-one years old, at the request of Gov. Lewis Cass, Dr. Zina Pitcher, and others, to lecture to the citizens of this young city about science. He then joined expeditions organized by Henry R. Schoolcraft for the expressed purpose of vaccinating Indians (Houghton was a physician). On these voyages, Houghton made extensive plant collections, finding several species new to science.

In 1839, Houghton, who was serving as state geologist, and George Bull, an assistant, discovered an unusual goldenrod at the north end of Lake Michigan, between Epaufette and Naubinway. The specimen of the goldenrod, along with several others, was sent to Asa Gray, who eventually recognized the goldenrod as a new species, naming it *Solidago houghtonii* in 1848. Ironically, Houghton had died three years earlier.

Houghton's goldenrod has a limited and unusual distribution. It occurs in interdunal hollows along the northern shores of Lake Michigan and Lake Huron at several localities in Michigan. There is also a single inland station for this species in Michigan, along the sandy shore of a lake in Crawford County.

More surprising, Houghton's goldenrod grows at a single locality in Genesee County, New York, where its habitat is a marly fen meadow. The New York colony is on land owned by a private conservation organization, but the plants are being threatened by the encroachment of the giant reed grass (*Phragmites australis*).

Solidago houghtonii is a rather distinctive perennial with a nearly smooth, often unbranched stem up to two feet tall. Five to fifteen yellow flowering heads are arranged in a nearly flat-topped cluster at the top of each stem. The flowers open during late summer.

Although this species has no federal status as yet, it is considered threatened in both Michigan and New York.

New England Violet or *Viola novae-angliae.*
Violet family. Perennial. Leaves basal, ovate, heart-shaped at base, hairy, toothed, to 2 inches broad. Flowers violet or purple, on leafless stalks; petals 5. Season: May, June.

It seems logical that the New England violet would be discussed in the chapter on plants of the northeastern states, but the majority of living colonies of this species occur in the Midwest.

The reason for the common name is that the first specimens ever found of this species were collected from Aroostook County,

Maine. Merritt Lyndon Fernald was exploring a sandy bank near Fort Kent on June 15, 1895, when he discovered this species. Three days later he found it again at the nearby community of St. Francis. All in all, the New England violet has been found at fourteen locations in Maine, including localities in the Penobscot Valley. It was thought to be restricted to a single station in 1981 in that state, but Susan Gawler of Maine's Critical Areas Program found five new but small populations during June 1982. It also occurs in adjacent New Brunswick, Canada. Its habitats in Maine include sandy banks of rivers and crevices of rocky ledges.

Viola novae-angliae has fared little better in New York, where it is apparently confined to Warren County. According to Richard Mitchell and Charles Sheviak's *Rare Plants of New York State*, the area in Warren County has been devastated year after year by ice rafting. The population was reduced to three plants by 1980, but Mitchell and Sheviak report finding about two dozen more in 1981, "in sandy and gravelly pockets among cobbles and other rocks along water courses."

No other New England locations, either historical or current, are known for this little violet. Several hundred miles to the west, however, is a second enclave of this species.

Dr. H. V. Ogden found the New England violet on a small sand island near Mercer, Wisconsin, in 1909, and a year later he collected it at Saxenville, in the same state. It was also in 1910 that W. S. Cooper discovered this violet on Isle Royale, Michigan.

Minnesota appears to be the home for most of the living colonies of *Viola novae-angliae*. Olga Lakela, whose pioneering work in Minnesota was followed in her later years by her study of the plants of south Florida, collected this violet in a hay meadow at Palo, Minnesota, in St. Louis County, in 1940. In 1941 she wrote that "it is one of the most common and showy violets that extends on the rocks from the north shore of Lake Superior to the Canadian boundary."

During this century, the New England violet has been found at several locations in northeastern Minnesota, northern Wisconsin, and northern Michigan. It occurs at four locations in four counties in Wisconsin today.

While I was studying endangered species in Michigan's Ottawa

National Forest during the summer of 1979, I tried without success to relocate the site of an early collection of this violet from dry ground near Ironwood in Gogebic County.

Viola novae-angliae is one of the stemless violets whose leaves and flower stalks arise directly from a short, thick, underground rhizome. The hairy flower stalk bears a rather large, purple or violet blossom in May and June. During the summer, a second type of flower, this one without petals, is formed on slender stems which arise from the rhizome.

Although the New England violet is rare throughout its range, except possibly in Minnesota, it has no federal status at the present time.

Sullivant's Sullivantia or *Sullivantia sullivantii*.
Saxifrage family. Perennial. Leaves basal, nearly round, toothed or shallowly lobed, long-stalked. Flowers small, borne on a leafless stalk to 1 foot; petals 5, white. Season: June to August.

When the state of Indiana is mentioned, it usually does not bring to mind rugged cliffs and waterfalls, heavily forested ravines, or slopes with hemlocks and pines. If you are a resident, or have visited the southern part of the state, however, you may know that these habitats are present.

Interstate Highway 64 bisects lower Indiana from west of Evansville to New Albany on the Ohio River. My wife and I have used this superhighway as a good central route from which to make forays to the north and to the south in southern Indiana.

On one such occasion we exited at St. Croix and headed a few miles north into the Hoosier National Forest. It was spring and we wanted to see the rich wildflower woods at a marvelously scenic area known as Hemlock Cliffs. From West Fork we followed a narrow gravel road to the parking lot at the top of the cliffs. A trail to the bottom of the ravine took us past toothworts and trilliums and several kinds of violets. The valley floor was aglow with masses of false rue anemone. On the cliffs ahead we spotted our first hemlock, and as we came closer to the end of the box canyon, several larger hemlocks lined the ridge above. Hemlocks in Indiana always surprise me (as well as make me envious), since that handsome conifer does not reach my native Illinois.

Just west of New Albany, we left I-64 one more time for some picturesque river bluffs several miles to the north. We were in search of the very rare plant with the unlikely name of Sullivant's sullivantia or, if you prefer the Latin name, *Sullivantia sullivantii.*

This small, cliff-inhabiting herb of the saxifrage family is known only from a few counties in southeastern Indiana, southern Ohio, and central Kentucky. The river along which the sullivantia occurs meanders back and forth between bluffs it had cut through centuries earlier. Along one south-facing exposure, the habitat was a dry slope with limestone shale exposed. Yellow chestnut oak, red oak, and shagbark hickory made interesting dominants among the trees on the slope. An attractive shrub which we identified later as the sweet viburnum (*Viburnum lentago*) grew beneath them.

Across the river, the moist, shaded, north-facing bluff was as different from the south-facing bluff as night from day. Although we saw no hemlocks in this area, the vegetation still was very exciting. A colony of the pale yellow-flowered grape honeysuckle (*Lonicera prolifera*), rare in Indiana, was growing along the bluff.

A tributary to the main river brought us to the sullivantia site. This species had been found here many years earlier on the shaded cliffs. Elsewhere in Indiana, it was known from a couple of rugged cliffs in Jennings and Clark counties.

The original collection of this species was made by William Starling Sullivant from limestone cliffs in Highland County, Ohio, in 1840. At this site, a rugged dolomite gorge capped with sandstone and with outcrops here and there of Ohio shale provides a refugium for plants unusual for this part of the country. Walter's violet, the Canada yew, and a rare, low-growing evergreen shrub known as cliff-green (*Pachistima canbyi*) grow here, along with the sullivantia. There is also a station for the sullivantia along a spectacular precipice in Hocking County, Ohio, where this species was collected in 1899 by W. A. Kellerman.

William Starling Sullivant, an Ohioan from the start, began his serious study of botany in central Ohio, publishing, in 1840, his *Catalogue of Plants, Native and Naturalized, in the Vicinity of Columbus, Ohio.* For a while Sullivant studied many kinds of plants, but soon he devoted himself almost exclusively to the investigation of mosses and liverworts. He became one of the greatest authorities on mosses the country has ever known.

Sullivantia sullivantii is a low-growing, slender perennial herb with several nearly round leaves at the base of the plant. A slender, erect stem no more than one foot tall bears a branched cluster of small white flowers at its upper end. The few known locations of this cliff-inhabiting species justify its review by the federal government for possible listing.

Iowa Golden Saxifrage or *Chrysosplenium iowense*.

Saxifrage family. Annual. Leaves alternate, simple, round-lobed. Flowers on 6-inch stalk, golden yellow, bell-shaped, 1/5 inch broad; stamens 5–8. Season: May to July.

Delicate might be a fitting adjective for the Iowa golden saxifrage (*Chrysosplenium iowense*), for its slender stem and slightly succulent leaves seem to be at a disadvantage among the many robust herbs and trees of the world.

This little species which grows on cool, moss-covered rocks was discovered by Prof. E. W. D. Holway on July 1, 1888, a few miles north of Decorah, Iowa. Although Per Axel Rydberg of the New York Botanical Garden described it as a new species thirteen years later, calling it *Chrysosplenium iowense*, botanists for half a century disagreed on the true identity of the plant.

In the meantime, additional colonies of this species had been found in Fayette, Dubuque, and Allamakee counties, all in Iowa, and at several locations in Alberta and Saskatchewan, Canada.

In 1947, Prof. C. O. Rosendahl of the University of Minnesota studied the golden saxifrages in detail, concluding that the specimens from Iowa and Alberta were referable to *Chrysosplenium iowense*.

Dr. Rosendahl found that the Iowa golden saxifrage had bright golden-yellow, shallowly bell-shaped flowers. In addition, Rosendahl discovered that the Iowa plants did not bloom the first year, but sent up erect, six-inch flower-bearing stems from the tips of the preceding year's stolons.

In recent years a colony of the Iowa golden saxifrage has been found in Minnesota. All of the Iowa and Minnesota locations are in the Driftless Area, that apparently unglaciated region surrounding the four corners of Iowa, Minnesota, Wisconsin, and Illinois. The habitat for the golden saxifrage is mostly cold-water seepage areas on north-facing wooded bluffs.

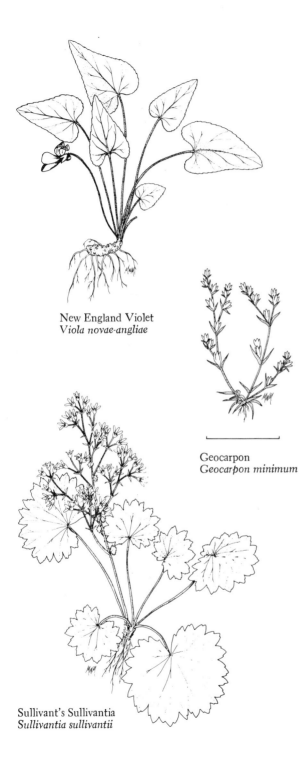

Iowa Golden Saxifrage
Chrysosplenium iowense

New England Violet
Viola novae-angliae

Geocarpon
Geocarpon minimum

Sullivant's Sullivantia
Sullivantia sullivantii

Whitlow Grass
Draba aprica

Since only five sites are known for the Iowa golden saxifrage in the United States, it deserves a place on the federal list of endangered and threatened species.

French's Shooting Star or *Dodecatheon frenchii.*

Primrose family. Perennial. Leaves basal, oval, long-stalked, toothless. Flowers few on a leafless stalk; petals 5, white or pink, pointing backward. Season: April, May. (See color plate 18.)

It was May 6, 1871. George Hazen French gathered his equipment together and pedaled away on his bicycle as he so often had done. He had his plant-collecting material with him, as well as an insect net and other essentials necessary to make good collections of butterflies, a specialty of his. And there was the indispensable lens which dangled on some braided twine from around his neck. In his bag was a lunch, for French would be gone all day.

George French was a naturalist and was on his way to a prospective new collecting site in southern Illinois. It was well into the morning when he arrived at the sandstone cliffs called Fern Rocks by the residents of the tiny nearby village of Makanda.

French decided to follow a sparkling rocky stream that paralleled the cliffs. Thousands of years ago this stream had cut the valley, which was teeming with spring wildflowers on that day. The forest floor was covered with blue and yellow and white violets, red wake robins and large white trilliums, and jack-in-the-pulpits and green dragons. There were gorgeous yellow celandine poppies, rich blue dwarf larkspurs, ground-hugging wild gingers whose flowers hid beneath their heart-shaped leaves, and the putty root orchid. The corrugated leaves of this orchid had withered a few days earlier, after having spent the winter in the forest nearly alone, because it was the only plant to show its green leaf during the winter months, except for occasional clumps of Christmas ferns and marginal shield ferns which overlooked it from the rock crevices of the nearby bluff.

As French made his way south through the wildflower-saturated woods, the sandstone cliff a few yards to the west kept pace, at some places towering nearly one hundred feet above, at others diminishing to a lesser height. At one place French decided to explore the bluff face where water constantly plummeted from the cliff's edge to a small shaded pool below. The bluff at this point was undercut a few feet westward, forming a picturesque and much shaded overhang. Indians in prior days no doubt sought shelter beneath overhangs like this, and these protected areas became known as shelter bluffs. The shelter bluff French found himself under was facing east at such an angle that the sunlight rarely penetrated through the forest canopy during any part of the day. Few plants could grow in such shaded conditions, except for a selection of mosses and liverworts and an occasional white-flowered bitter cress. But there was something else this time.

French's eyes fell upon several rosettes of light green leaves growing in the wet soil beneath the overhang. From the center of each rosette was a flowering stalk with five or six blossoms. French recognized the flowers as shooting stars, but the leaves weren't like those of the shooting stars growing on the bluff tops overhead. Yet the flowers, although somewhat smaller, seemed identical to those of the common shooting star (*Dodecatheon meadii*). With a few specimens of this unusual shooting star in his bag, French was on his way again.

Unable to satisfy himself as to the plant's identity, French finally sent a specimen to Dr. George Vasey, an Illinois friend and colleague who had been named curator of the United States National Herbarium in Washington, D.C.

Recognizing that the plant resembled the common shooting star but differed in leaf shape and habitat, Vasey named it after its collector, calling it *Dodecatheon meadii* var. *frenchii* in 1876.

French and botanists after him found more colonies of this new shooting star in southern Illinois, always under sandstone overhangs. In 1932, when French was ninety years old and long retired as a biology professor at Southern Illinois University in Carbondale, Dr. Per Axel Rydberg of the New York Botanical Garden came upon French's old collections and suggested that they really represented a new species. Accordingly, Rydberg changed the name of this plant to *Dodecatheon frenchii*, giving it full species status.

Botanists were divided over the correct identity of the plant. Some sided with Rydberg; others felt that the difference in leaf shape was in some way due to the shaded environment, and that French's shooting star was merely a variety of the common shooting star.

In 1950, Prof. John W. Voigt of Southern Illinois University transplanted specimens of French's shooting star to open bluff tops, and took the common shooting star from the ridge top and planted it below the sandstone overhang. After several years, French's shooting star and its offspring had remained constant, while the common shooting star was unable to survive the deep shaded conditions. Fifteen years later, Prof. Leslie Olah of the same university studied the cytology of these two shooting stars and found that they had different numbers of chromosomes in their cells. The evidence was in. French's shooting star was a good species.

During the decade of the 1970s, French's shooting star was discovered at one locality in northwestern Arkansas, one in southeastern Missouri, one in southern Indiana, and a few in Kentucky.

The telltale leaves of French's shooting star taper abruptly to a well-defined leaf stalk. (The common shooting star has no leaf stalks.) White- or occasionally lavender-flowered blossoms are like those of the common shooting star.

Despite the fact that French's shooting star has been found in a number of locations, all colonies are in a very restricted habitat. As a result, this species is being reviewed by the United States Fish and Wildlife Service for possible federal listing.

Synandra or *Synandra hispidula.*
Mint family. Perennial; stems to 18 inches, hairy. Leaves opposite, simple, toothed, heart-shaped at base. Flowers to 1½ inches, several in a terminal spike; petals 2-lipped, white with lavender or purple stripes. Season: April, May. (See color plate 19.)

Among the many plants discovered by André Michaux during his twelve-year stay in the United States is the lovely and delicate synandra, a large, white-flowered member of the mint family. Michaux discovered it in the forests of Tennessee during one of his forays into the hinterland of the eastern United States.

Michaux arrived in New York in 1785 with his fifteen-year-old son François and his gardener, Saunier, at the expense of the French government. His mission was to find and send back to his mother country plants of medicinal, agricultural, or timber potential. He established a garden in New York, which he left in the care of Saunier while he ventured south to Charleston, South

Carolina. Here Michaux established another garden and began his exploration for plants that would take him to the Bahamas; Florida; the mountains of North Carolina, Tennessee, and Kentucky; the forests of Pennsylvania, New Jersey, and New York; and the wilderness of Indiana and Illinois. His travels led him as far west as Fort Kaskaskia on the banks of the Mississippi River in southern Illinois.

As he was returning to France in 1796 during the aftermath of the French Revolution, his boat shipwrecked and he was knocked unconscious. When he finally arrived in Paris, he found that many of the plants that he had been shipping had been destroyed and that the gardens and most of the friends that he had known had disappeared. The new French government would not allow Michaux to return to the United States, where he wished to go, but it did permit him to join an expedition to Australia. On his way, he stopped to explore the Mascarene Islands. He contracted a fever and died in Madagascar in 1802. His *Flora Boreali-Americana* was published posthumously.

When Michaux found synandra, he was reminded of the European genus *Lamium*. Accordingly, he named the plant *Lamium hispidulum*, but after Thomas Nuttall described the new genus *Synandra* in 1818, Michaux's plant eventually became known as *Synandra hispidula*, the binomial it retains today.

The lack of a common name, other than synandra, attests to the rarity of this species. Since its discovery, synandra has been found sparingly in Virginia, West Virginia, North Carolina, Alabama, Tennessee, Kentucky, Ohio, Indiana, and Illinois. In some of these states synandra is found at a number of sites, but in other states it is exceedingly rare. It grows in North Carolina on rich wooded slopes in Swain County, while in Alabama it is found only in a rich, rocky, limestone woods in Jackson County.

There are four Illinois locations, all in or near Jackson County in southwestern Illinois, where it grows in deep, rich, mesophytic forests. I saw my first synandra in 1953 in the exact location that Professor George Hazen French had discovered it in Illinois in 1871. The sight was one to behold, and should be seen by every wildflower enthusiast in the country.

The deeply shaded woods in which synandra was growing slopes gently to a wet, east-facing sandstone cliff. It was during the first week of May, and the forest floor was so heavily carpeted with a luxurious growth of spring wildflowers that it was difficult to leave the main trail, for every footstep into the forest meant trampling specimens of violets, wild gingers, wild geraniums, dutchman's breeches and squirrel corn, waterleafs and phacelias, and jack-in-the-pulpits.

The synandra occupied an area of about 200 square feet, where its large lavender-streaked white flowers contrasted beautifully with the delicate blossoms of the pink valerian (*Valeriana pauciflora*), the pert and saucy flowers of blue-eyed Mary (*Collinsia verna*), and the majestically pure white flowers of the giant trillium (*Trillium flexipes*). Every spring I make a pilgrimage back to that same wooded slope, where I can renew my faith in the natural wonders of the world and feel peace in my heart.

Synandra hispidula stands about fifteen to eighteen inches tall, with a hairy stem. The lavender-striped white flowers occupy the upper one-fourth of the plant in an unbranched spike.

There are too many known locations in synandra's total range to justify a federal listing for it, but several states where it is very rare list it as endangered.

Mead's Milkweed or *Asclepias meadii*.

Milkweed family. Perennial; stems to 2 feet, smooth, containing milky sap. Leaves opposite, simple, broadly lance-shaped, toothless, without stalks. Flowers borne in umbrellalike clusters; petals 5, green or greenish white; hoods purple or greenish purple. Season: May, June. (See color plate 20.)

In 1856 the celebrated American botanist John Torrey described a milkweed which had been found a short time earlier by Dr. Samuel B. Mead near Augusta, in west-central Illinois. This new milkweed, named *Asclepias meadii* for its discoverer, was an inhabitant of prairies.

Neither Torrey nor Mead was trained as a botanist. Torrey was a physician and practiced medicine for a while. Later, he taught chemistry and then served as assayer for the United States Assay Office. All the time, he pursued his favorite subject of botany, collecting plants and writing his *Catalogue of North American Flora* in 1819 and his *Compendium of the Flora of the Northern and Middle States* in 1826. This led to his grandiose plan to write a complete flora of North America. He invited his protégé, Asa

Gray, to collaborate on the work. This dynamic duo completed several volumes of the project before it had to be abandoned because of other commitments. Nonetheless, Torrey and Gray's flora stands as a monument to nineteenth-century botany.

Samuel B. Mead also was a physician, having obtained his medical degree from Yale in 1824. From 1834 to 1880, while practicing medicine in Augusta, Illinois, Mead collected several thousand plant specimens from Illinois. He was Illinois's first resident plant collector, and in 1846 he published the first catalogue of Illinois plants.

Shortly after Mead's original collection, this milkweed was found scattered in four additional counties in the northern half of Illinois; in adjacent Iowa, southeastern Wisconsin, and northwestern Indiana; and across Missouri to eastern Kansas. At no locality, however, could the species be considered abundant.

During the last half of the nineteenth century, America's population migrated westward. Most of these pioneers were tillers of the soil. When they discovered that the black soil which supported prairie vegetation was ideal for agricultural crops, the prairies were plowed and replanted to corn and wheat. One of mid-America's great natural habitats was reduced overnight, and with it went the prairie grasses and the prairie wildflowers. The Wisconsin colony of Mead's milkweed disappeared as early as 1879, and the Indiana plants shortly afterward. In Illinois it was extirpated from all five of the counties where it had originally been found, but a new locality was discovered in a prairie next to a railroad in northeastern Illinois during the 1970s. It was found at one location in Iowa after it was reportedly extirpated from that state.

In Missouri, the state where Mead's milkweed has been found most often, it still is scattered in a few counties, primarily in the west-central part of the state. There are also a few locations for this plant in eastern Kansas.

At all of these localities, past and present, *Asclepias meadii* grows in prairies or on rocky glades. But an unusual habitat for this species exists in southern Illinois, a region more than 200 miles from the nearest population.

I was exploring a picturesque area of the Shawnee National Forest southeast of Harrisburg, Illinois, during the latter part of May 1953, when I came upon two plants of a milkweed unknown to me. I was at the edge of a sandstone escarpment, where a shallow layer of soil supported sparse vegetation. Nearby, scrubby post oaks and blackjack oaks were struggling to survive in the hot and arid habitat. The milkweeds, in flower at the time, were identified as *Asclepias meadii*, with both the habitat and the extreme southern locality of the plants coming as a great surprise. Incidentally, two plants of this species still exist at the same site, nearly thirty years later. In 1981, another plant was found under similar conditions on a sandstone cliff a few miles away.

Asclepias meadii is a perennial herb growing to a height of about two feet. Several fragrant greenish or greenish-purple flowers are arranged in an umbrellalike cluster at the tip of the stem. From the cluster of flowers are produced one or two milkweed pods that contain hairy-tufted seeds.

The future existence of Mead's milkweed is a troubled one. Because of this, it should be added to the federal list of endangered species.

Kankakee Mallow or *Iliamna remota*.

Mallow family. Perennial; stems hairy, to 6 feet. Leaves simple, alternate, lobed and toothed. Flowers pale pink; petals 5, 1 inch long. Season: August, September. (See color plate 21.)

It seems rather unlikely that a small island in the Kankakee River of northeastern Illinois would be the home of one of the rarest flowering plants in the country. But that is precisely where the Kankakee mallow (*Iliamna remota*) lives.

This handsome hollyhocklike plant was discovered by the Rev. E. J. Hill while he was employed as a teacher at the Kankakee high school in 1872. During the summer of that year, Reverend Hill made his way to the island across from the village of Altorf, about nine miles north of Kankakee.

The plant Hill found was a coarse perennial with its branched, hairy stems standing about six feet tall. Several large, pink, slightly fragrant flowers were formed in an inflorescence that terminated each branch.

Hill's specimens were erroneously identified at first as *Sphaeralcea acerifolia*, a mallow of the western United States. It wasn't until 1899, more than a quarter of a century later, that Dr. Edward L. Greene revisited the island in the Kankakee River and came to the conclusion that the mallow growing there was a new species. Greene called it *Iliamna remota* because of the great distance between it and its nearest relative.

In 1912, while preparing the second edition of *Britton and Brown's Illustrated Flora of the Northern United States and Canada*, Nathaniel Lord Britton sought additional specimens of the Kankakee mallow. Obligingly, Dr. Jesse M. Greenman and Dr. Earl Edward Sherff of the Field Museum of Natural History in Chicago accompanied Reverend Hill, then seventy-nine years old, to the island in the Kankakee River. Sherff's account of the foray is worth retelling.

> Arriving in the forenoon of August 3rd at Altorf, just northeast of the island's upper end, we forded the river at a point somewhat upstream, using a horse-drawn carriage. Securing a boat on the opposite shore, we rowed to the southwestern shore of the island and anchored. We were led presently by Mr. Hill with surprising directness and accuracy to the very spot where he remembered having collected some forty years before. The plants of *Iliamna remota* . . . were numerous.

A few plants were dug up and transplanted to gardens, and numerous seeds were gathered and sent to botanists in many parts of the world. To this day, the Kankakee mallow still grows on the same island where Illinois botanists consider it an endangered species.

Additional factors complicate the story, however. In 1944, Dr. S. W. Witmer of Goshen College discovered four colonies of the Kankakee mallow along the north side of the Wabash Railroad right-of-way near New Paris, Indiana, at a point where the railroad crosses the Elkhart River. It has been assumed that these plants are escapees from cultivation or were intentionally sown as a method to perpetuate the species.

Meanwhile, Dr. Earl Core of West Virginia University, while exploring the summit of Peters Mountain, Virginia, in 1927, found a plant nearly identical to the Kankakee mallow. Surely this montane plant could not be the same thing that was growing wild in the Kankakee River of northeastern Illinois, several hundred miles distant. Yet the only apparent differences were that the Peters Mountain plants were shorter, their flowers lacked an odor, and the leaf lobes were more oblong. Because of these few differences, the different habitat, and the distant location, the Virginia plant was named *Iliamna corei*.

Some botanists today believe that the Kankakee mallow and the Peters Mountain mallow are distinct species. They are considered so by Virginia botanists in their *Endangered and Threatened Plants and Animals of Virginia*, where they list *Iliamna corei* as an endangered species.

Other botanists think that the Peters Mountain plant is a variety of the Kankakee mallow. Still others believe the two are really one and the same. To add to the fun, two colonies of an *Iliamna*, seemingly close in appearance to the one from Illinois, have been found along a railroad and a highway in two mountain counties of western Virginia.

Whatever the ultimate decision is about the identity of all of these mallows, they are rare enough to justify consideration as a federally endangered species.

Geocarpon or *Geocarpon minimum*.

Carnation family. Annual; stems less than 2 inches. Leaves opposite, simple, crowded near base of plant, to ¼ inch. Flowers less than ¼ inch broad; sepals 5, green or purple; petals none. Season: May.

Few states have had their plant life catalogued more thoroughly than Missouri. Benjamin Franklin Bush in the latter part of the nineteenth century and Ernest Jesse Palmer during the first four decades of the twentieth century laid a solid foundation for a Missouri flora. However, it was a young botanist from St. Louis, Julian A. Steyermark, who explored every corner of Missouri for more than a quarter of a century, beginning in 1930, to obtain information for his epic *Flora of Missouri*. (After completing his great Missouri work in 1963, Steyermark migrated to Venezuela, where he has discovered hundreds of species new to science and even had a major waterfall named for him. He resides in Caracas today.)

It was my great fortune that Dr. Steyermark was a visiting professor in botany at Southern Illinois University during my first year on the faculty at that institution. We had several things in common. We both had done botanical research at the Missouri Botanical Garden (Shaw's Garden) while pursuing advanced degrees at Washington University in St. Louis. We were both writing floras. He was completing his *Flora of Missouri*, and I was embarking on my *Illustrated Flora of Illinois*.

Dr. Steyermark, Prof. John W. Voigt, the department ecologist, and I shared a sixteen-by-eight-foot office for the three-month spring quarter in 1958 at Southern Illinois University.

From mid-March to late April, Steyermark enthralled me with stories of his experiences. I was particularly excited about his rediscovery in 1957 of the very rare and little known *Geocarpon minimum*. This species had been found originally in Jasper County, Missouri, in 1913 by E. J. Palmer. This remained the only known location until Steyermark rediscovered it in St. Clair County in 1957. My nagging at Dr. Steyermark to show me geocarpon paid off. He agreed to take Dr. Voigt and me with him on a three-day excursion to see if we could discover additional stations for this dwarf rarity.

We left Carbondale on May 2, heading in the general direction of Springfield, Missouri. Leaving Cape Girardeau, we stopped briefly at the old Bollinger Mill and then headed into the Ozarks. Up and down county highways, past shortleaf pine woods, over clear, rock-bottomed streams we went, Steyermark pointing out species right and left. We made frequent stops, usually after an enthusiastic comment such as, "Wait! There's a county record." I tried to absorb all the species as fast as I could, but Steyermark never relented. We screeched to a halt in Bollinger County, pulling off the road next to a steep, wooded, cherty slope. Julian asked if either of us had ever seen the little rare sedge, *Scirpus verecundus*. We replied in the negative, and off we went down the slope. *Scirpus verecundus* was not known from Bollinger County, but Steyermark was optimistic, saying that this slope looked just right for the little plant. In less than five minutes the woods echoed Steyermark's voice: "Whee! Over here. I've found it." Scrambling over the treacherous chert, Voigt and I made it to the kneeling botanist, who crouched over a four-inch-tall tuft of green. I had seen my first *Scirpus verecundus*, and I was never to forget the habitat. Some fifteen years later, while exploring in extreme southern Illinois, I came upon a similar steep, cherty slope. Recalling my earlier experience, I began to comb the woods and was rewarded by making the first discovery of *Scirpus verecundus* in Illinois.

We persisted beyond supper and into the duskiness of twilight. By dark, we were still on the road. I began to understand how Steyermark had been able to canvass every square foot of Missouri in thirty years—he never stopped to rest. I saw my first Ozark tarantula, scurrying across the road in the glow of our headlights. Finally we stopped for camp at about nine-thirty, reeling from the vast amount of information imparted by Dr. Steyermark.

We arose early on May 3 in anticipation of seeing geocarpon. We had entered Polk County near Graydon Springs when Dr. Steyermark announced that we had arrived at a likely-looking spot for geocarpon. It was a glade on a west-facing sandstone escarpment next to Coates Branch, a tributary of the Little Sac River. We quickly found Butler's quillwort (*Isoetes butleri*), flower-of-an-hour (*Talinum parviflorum*), and sandwort (*Arenaria patula*), species known to be associated with geocarpon at its other locations. Then, in a small, moist depression of the glade, we spotted it—dwarf, two inches tall, semisucculent. At this time of year the plants were wine-purple in color. Once we knew what we were looking for, we began to spot another, and another, and yet another. Our discovery marked the third known location for geocarpon in the world! On the same glade I also saw for the first time in my life the golden selenia (*Selenia aurea*) of the mustard family and the Texas saxifrage (*Saxifraga texana*).

Our spirits were buoyed by these discoveries. We hastened into Greene County and stopped at a likely looking sandstone glade near Pearl. We noticed the golden selenia and Texas saxifrage from a distance and, more closely, Butler's quillwort. It surely would be just a matter of time. Sure enough, the wine-purple plants of *Geocarpon minimum* became evident after we had wandered into an adjacent glade.

As we were hopping over the glade in our crouched positions, we happened upon a small plant previously unknown from Missouri. It was *Scleranthus annuus*, the awlwort of the pink family.

With our success in finding geocarpon, we headed into Dade and Cedar counties, stopping at several glades. Much to our consternation, we wouldn't find geocarpon at any of them. By late afternoon, we started back to the east, where we had planned to camp at beautiful Alley Springs. Darkness caught us again while still on the rolling hills of the Ozarks. The night scarcely slowed up Dr. Steyermark as he continued to name roadside plants as our headlights struck them. "*Andropogon elliottii*," he shouted once, as we whizzed by a clump of two-foot-tall grasses. Most people have trouble identifying this species in the daylight with the specimen in front of them, but Steyermark's identification was right on the money.

Following a beautiful night under Ozark skies, we spent all day Sunday looking for plants as we made our way home. We got back to Carbondale early in the evening, finishing a three-day red-letter trip I will never forget.

Geocarpon minimum was so poorly known for nearly half a century after its initial discovery by Palmer that it was even placed in the wrong family. Steyermark's studies showed it to be a member of the Caryophyllaceae, or carnation family, rather than the Aizoaceae, or carpet-weed family. This smooth little annual has slender branched stems less than two inches tall.

At about the time that Steyermark was finding his new locations for geocarpon in southwestern Missouri, Dr. Dwight M. Moore of the University of Arkansas found it on sandy barrens at two locations in northwestern Arkansas.

This species should be considered strongly for federal listing because of its extremely limited range.

Whitlow Grass or *Draba aprica*.

Mustard family. Annual; stems to 1 foot, slender. Some leaves crowded near base of plant, oval, few-toothed; leaves on stem alternate, narrower, toothless. Flowers in dense clusters; sepals 4, green; petals 4, white, 1/10 inch long; stamens 6. Pods slender, to ¼ inch. Season: April.

One of the most distinctive topographic features of the midwestern United States is the Ozark Mountains of Missouri and Arkansas. This mountain system is extremely old and known for its unique and diverse flora. Scenic areas abound in the Ozarks and attract great numbers of summer recreationists for camping, hiking, and canoeing.

The Ozarks are characterized by rough, rocky mountains of granite, dolomite, and sandstone, alternating with deep, heavily wooded ravines. Occasional treeless glades punctuate the forest. Perhaps the greatest attractions in the Ozarks are the clear mountain streams which flow rapidly over a gravelly and rocky bottom.

Rugged boulders which have fallen to the streams from nearby cliffs add to the spectacular nature of the streams. Areas such as Johnson's Shut-ins and Jam-up Bluff typify the region.

As I like to do when visiting the Ozarks, I was hiking along a heavily graveled stream in southern Missouri one day in late April. Among the plants that live along and sometimes in the clear streams are some which grow only in the Ozarks. One of these is a small tree which puts forth its bright yellow blossoms in January and February, often when snow is on the ground. This is the vernal witch hazel, *Hamamelis vernalis*, and it was just beginning to form its woody capsules that spring day. Another plant which would bloom later in the spring had just come up, barely showing its leaves, which would be glossy in about a month beneath the cluster of blue flowers. This was the shiny-leaved Ozark blue-star, *Amsonia illustris*, a species confined to the Ozarks.

As I follow the Ozark streams, I often take short side trips up a ravine or onto an inviting glade or over to a sheer rock face. On this occasion I saw a low gravelly mound some twenty feet off the stream bed and decided to investigate. Much to my surprise, several low, slender plants with small white flowers were in full bloom.

Obviously the plant was a member of the mustard family because of its four white petals and six stamens. I turned to that family in my field-worn copy of *Flora of Missouri* by Julian A. Steyermark and found the plant to be a whitlow grass (*Draba aprica*).

Steyermark's manual states: "occurs in low rocky or alluvial woodlands in valleys of the southeastern Ozark section. Ranges from Georgia to Arkansas and southeast Missouri. This is a very rare species and is known from only a few stations in the United States."

Thomas Nuttall first discovered this rare little mustard somewhere in Arkansas in 1819. His original collection went unnoticed and unnamed until 1838. In that year, John Torrey and Asa Gray described it as a variety of *Draba brachycarpa*. From 1838 to 1901, this plant was virtually forgotten.

Then, on May 9, 1901, the southern botanist C. B. Beadle, exploring on Kennesaw Mountain near Marietta, Georgia, found this same mustard. Unable to identify the plant, Beadle sent it to his colleague, Dr. John Kunkel Small, who believed it was a new species and named it *Draba aprica*.

It wasn't until Merritt Lyndon Fernald of Harvard University undertook a detailed study of the genus *Draba* in North America that it became apparent that Nuttall's plant from Arkansas was the same thing as Beadle's plant from Kennesaw Mountain. Fernald agreed with Small that it was a good species, and retained the name *Draba aprica*.

Dr. Steyermark made his first collection of this species from Missouri in 1930. It has since been found in half a dozen areas in the Missouri Ozarks. Although it sometimes grows on gravelly mounds near streams, it also is found on steep oak-pine slopes, in low rocky woodlands, and in dry rocky ground.

In 1957, Dr. Reed C. Rollins, Gray Professor of Botany at Harvard University, and Kenton Chambers found *Draba aprica* north of Broken Bow, Oklahoma. A surprising discovery of this little mustard occurred on April 12, 1969, when S. W. Leonard and D. Hobbs found it in Lancaster County, South Carolina.

Draba aprica is an annual whose slender stems grow less than one foot tall. The white flowers are borne in dense clusters. The fruit is a slender pod about one-fourth inch long and covered by microscopic star-shaped hairs.

Of the eleven localities where this species has been found, the Arkansas site has not been relocated since Nuttall's record in 1819, and one Missouri station was destroyed when the Black River was dammed to form Clearwater Lake. Clearly, this small mustard is in need of federal protection.

❧ 5 ❧

Fleeting Flora

of the South-Central States

The south-central states in this book include Arkansas, Louisiana, Oklahoma, and Texas. Ten plants from these states are either on the federal list or are currently being considered for listing. Of these, seven are cacti. The McKittrick pennyroyal is included in this chapter, although it also occurs in New Mexico.

Texas Wild Rice or *Zizania texana.*
Grass family. Aquatic perennial; stems to 8 feet. Leaves elongated, to 3 inches. Flowers inconspicuous, arranged in 1-foot-long clusters; male and female flowers on separate branches. Season: April to November.

By the time summer of 1982 rolled around, only three grasses had made the federal list of endangered and threatened species, despite the fact that there are several very rare grasses among the hundreds that grow in the United States.

One of those grasses was placed on the federal list on April 26, 1978. It is the Texas wild rice (*Zizania texana*), a large aquatic grass confined to a very small area of the San Marcos River in Texas.

Although the Texas wild rice had been collected at least twice between 1892 and 1921, it was not until 1932 that an astute amateur botanist from San Antonio, W. A. Silveus, saw this plant in the San Marcos River and knew it was different from any previously named wild rice. The two early collections had been referred to *Zizania aquatica*, a species known from New England to Wisconsin and from the south Atlantic coast to Louisiana.

Silveus, an attorney who published his important book *Texas*

Grasses in 1933, had corresponded with Agnes Chase, the noted grass authority at the United States National Herbarium in Washington, D. C., in April 1932 about the unusual *Zizania* he had found. In his letter, Silveus wrote that the "grass was growing in water from 1–4 ft. deep mostly some distance from the bank— the plant prostrate on or just under the surface of the water. . . . I could understand if the river should rise suddenly and cause this grass to become prostrate on the water, but this stream has a pretty steady flow all the time. . . ." In another letter to Mrs. Chase that year, Silveus wrote that the grass "blooms from April to Nov. anyway and the man at the pump station on the bank of the lake says it blooms all year if warm. He says they clear it out and it comes up again immediately. It covers several acres on lake and along stream below."

The lake referred to is Spring Lake, which was created early in the twentieth century by damming the San Marcos River near the source of the springs. These springs discharge unusually clear water at the rate of nearly 200 million gallons daily. The San Marcos River at its upper end flows swiftly over alternating shallow and deep pools in a channel up to fifty feet wide and almost twelve feet deep. The springs which feed the river from their limestone source keep the slightly alkaline to neutral water at a nearly constant temperature.

Silveus indicated that at the time he discovered the grass, it was so abundant in the San Marcos River, in Spring Lake, and in nearby irrigation ditches, that the irrigation company had to work continuously to keep the growth of the wild rice under control.

In 1967, Dr. William Emery of nearby Southwest Texas State University, who has spent several years studying this species, reported that only one plant of *Zizania texana* occurred in Spring Lake, clinging tenuously to the channel bank directly behind the sluiceway to the dam, and that there were no plants of this grass in the first half mile of river below the dam. Beyond that, the plants were scattered over a distance of one and one-half miles. At that time Emery discussed the factors going against the Texas wild rice. The upper six to eight feet of vegetation in the lake were cut periodically by an underwater mowing machine to keep the vegetation from clogging up so that tourists would have a more attractive lake. (A private enterprise has begun offering glass-bottom boat rides on the lake.) After the mowing, great masses of floating debris moved rapidly downstream, passing over the wild rice that was growing in the river. This periodic rush of debris over the grass apparently prevented pollination from occurring, and the plants failed to produce viable seeds. In addition, since the San Marcos River where the grass occurs flows through the city, city workers continuously plowed or harrowed the river bottom. Accidental discharge of untreated sewage sometimes entered the river. An unusual factor affecting the habitat was the operation of a private aquatic plant business where the workers from the company would remove from the river plants which had no commercial value and replace them with plants which could be sold. Not only did this rid the river of its native vegetation, it created a constant source of turbidity.

Ten years later, Emery reported that all of these adverse elements had abated, but the Texas wild rice was now eliminated from Spring Lake and occurred only in an area of 1,131 square meters of river habitat. The plants that remained were still unable to produce seeds.

Emery has conducted several interesting experiments with this grass. After he had transplanted some to a spring-fed sluice with constant temperature on the campus of Southwest Texas State University, he noted that the stems and leaves of the grass, which lay in the water in the wild populations, grew erect and emerged well above the water. In addition, pollination occurred and several hundred seeds were produced. Emery was able to germinate several seedlings, and after they had obtained sufficient size, he reintroduced some of them into Spring Lake. However, they were eaten virtually overnight by nutrias which live in the area.

Texas wild rice, which grows in large clumps that are rooted in the river bottom, has stems and leaves that are immersed in the fast-flowing water. Although the plants seldom flower, when they do they have their flower clusters somewhat elevated above the water.

Texas Poppy Mallow or *Callirhoe scabriuscula*.

Mallow family. Annual; stems to 4 feet, hairy. Leaves alternate, simple, deeply 3- to 5-lobed. Flowers showy, wine-purple, cup-shaped; sepals 5, green; petals 5, 1 inch or more long. Season: May, June. (See color plate 22.)

One of the most beautiful groups of wildflowers that are exclusively North American is the genus *Callirhoe* of the mallow family. Called poppy mallows, these plants have showy white to purple flowers which superficially resemble poppies, but which structurally have all the characteristics of the mallow family.

Rarest of the poppy mallows is *Callirhoe scabriuscula*, known as the Texas poppy mallow. The original collection of this species was made by Sutton Hayes, an army surgeon with the El Paso and Fort Yuma Wagon Road Expedition during the latter half of the nineteenth century. Hayes's specimen, which was found somewhere along the Colorado River in Texas, made its way to the Gray Herbarium of Harvard University, where Benjamin L. Robinson named it *Callirhoe scabriuscula* in 1897.

For half a century the Texas poppy mallow was known only from the single specimen collected by Hayes. Then, on June 1, 1942, a plant explorer named Cory rediscovered this species in Runnels County in west-central Texas.

Today, *Callirhoe scabriuscula* is confined to a single county, where it lives in a small area of deep sandy soil blown from alluvial deposits along the Colorado River.

There are several threats to the already limited numbers of plants of the Texas poppy mallow. It is subjected to some grazing, and observers have noted that when grazing occurs, the species declines. There is a very serious threat from sand-mining operations, since the Texas poppy mallow lives in deep, sandy soils. Road construction apparently has eliminated some specimens. Of special concern is the threat from gardeners who might transplant this species to gardens, since the wine-purple flowers are so handsome that the Garden Club of America calls the poppy mallow one of the most beautiful wildflowers in Texas.

Callirhoe scabriuscula is an annual that stands two to four feet tall. Its five magnificent wine-purple petals form an open, cup-shaped flower that blooms in May and June.

Since there is imminent threat from several sources to this already rare species, it was listed as a federally endangered species on January 13, 1981.

Navasota Ladies'-Tresses or *Spiranthes parksii.*
Orchid family. Perennial. Leaves long and narrow, crowded near base of plant, without teeth, absent at flowering time. Flowers

on 1-foot leafless stalks, borne in a spiral or twisted spike, white; lip petal to ¼ inch long. Season: September, October.

H. B. Parks was exploring in a post oak woods adjacent to the Navasota River in east-central Texas in 1945 when he discovered a slender, foot-tall plant with small white orchid flowers produced in a spiral pattern. The plant was obviously some kind of a ladies'-tresses orchid, and it was identified as one of several species of this genus known from Texas. But two years later, a fellow Texan, Dr. Donovan S. Correll, examined the specimen that Parks had collected and decided it was a new species, calling it *Spiranthes parksii.*

For thirty years no one was able to relocate this orchid in the wild, and suggestions were being made that it was probably just an abnormal form of one of the three other kinds of ladies'-tresses that were growing nearby.

Then, in 1978, P. M. Catling rediscovered *Spiranthes parksii* from Brazos County, Texas, the same county in which Parks had made his early collection. Before the excitement of Catling's rediscovery had waned, a second population was found, presumably near the same place that the plant was originally found in 1945. Today there are fewer than twenty known plants of this orchid.

The two recently found populations of the Navasota ladies'-tresses are not without some endangerment to their existence. Both are on private property. One is growing adjacent to an urban area, and the other is on land used primarily for hunting.

Spiranthes is a large genus of over 300 species, almost all of which are found in temperate or tropical regions of the New World. The generic name is derived from the twisted, or spiral, manner in which the flowers are arranged in the spike. This character also is responsible for the common name ladies'-tresses, because of the vague resemblance of the flower spike to braided hair.

Spiranthes parksii stands about one foot tall. As is the case with all orchids, it is the lip petal of the flower that provides the distinguishing characteristics for this species. The lip is white, with a green stripe down the center and shallow teeth around the outer edge. The leaves are absent at flowering time.

Because of the very limited occurrence of this species, it was declared a federally endangered plant on May 6, 1982.

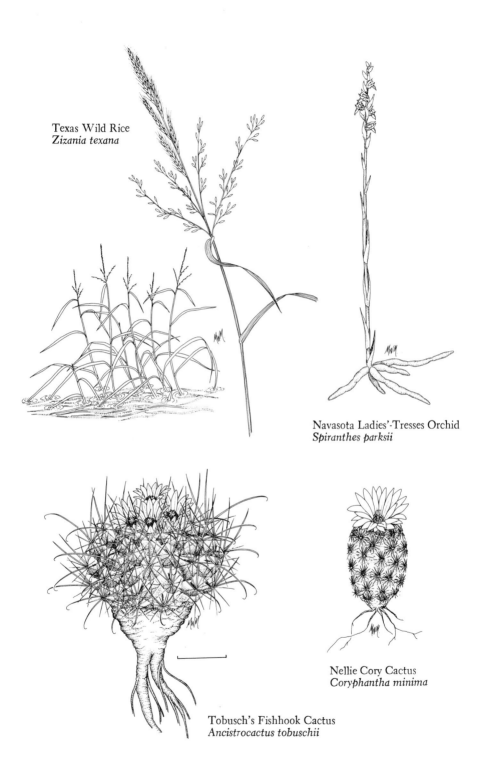

Texas Wild Rice
Zizania texana

Navasota Ladies'-Tresses Orchid
Spiranthes parksii

Tobusch's Fishhook Cactus
Ancistrocactus tobuschii

Nellie Cory Cactus
Coryphantha minima

Tobusch's Fishhook Cactus or *Ancistrocactus tobuschii.*
*Cactus family. Succulent; stems usually solitary, top-shaped, to
3½ inches in diameter, with 8 ribs; central spines 3, yellow with
a red tip, to 1½ inches, one of them hooked at tip; radial spines
7–12, straight, slightly shorter than centrals. Flowers yellow, to
2 inches across. Season: February to June.*

One of the most distinctive physiographic regions in Texas is the
Edwards Plateau, considered by most geographers to be a south-
eastward extension of the Great Plains, although its characteristics
and features are quite different. The Edwards Plateau lies south
of the Llano Estacado, a high-elevation grassy area usually referred to as the Staked Plains. It is bounded on the west by the
Trans-Pecos Mountains. To the southeast, the Edwards Plateau
extends between the Colorado and Rio Grande rivers as far as the
Balcones, an escarpment between Del Rio and Austin.

Elevation within the Edwards Plateau ranges from 800 feet
above sea level at the Balcones to about 3,500 feet at the base of
the Davis Mountains north of Big Bend National Park.

Most of the plateau is underlain by spongy Edwards limestone,
which provides an important reservoir of water for San Antonio
and other nearby cities. Some notable springs flow in this area.
Mesquite and native grasses form a great part of the vegetation
on the plateau, and provide food for large herds of livestock.

Among the many interesting aspects of the Edwards Plateau
are several kinds of plants and animals that are generally restricted to this area. One of these is a cactus that grows only in a
thirty-mile-long strip west of San Antonio, as well as in one disjunct locality in Big Bend National Park.

The cactus is known as *Ancistrocactus tobuschii* and was discovered by H. Tobusch in 1951. It grows in rugged canyons which
cut into the edge of the Edwards Plateau in Bandera and Kerr
counties. The original discovery was made in an area of limestone dominated by junipers, oaks, and various grasses at an elevation of about 1,400 feet.

A year after Tobusch's discovery, this little cactus was found
in the same general locality by W. T. Marshall and E. R. Blakley
and, a decade later, by Del Weniger. Weniger, a cactus enthusiast
who wrote the beautifully illustrated *Cacti of the Southwest*, com-
ments that Tobusch's fishhook cactus is one of the rarest forms in

the Southwest. He estimated in 1969 that the population of *Ancistrocactus tobuschii* in the Edwards Plateau country did not comprise more than a few hundred plants in all.

One of the most surprising and unexpected discoveries of Tobusch's fishhook cactus was made by Roland H. Wauer during the 1970s. Wauer, a naturalist at Big Bend National Park, was searching for cacti in the park when he discovered this rare species growing in the mountains. The Big Bend plants are about 250 miles from the Edwards Plateau populations.

When the United States Fish and Wildlife Service listed this species as federally endangered on November 7, 1979, it noted that there were fewer than 200 specimens remaining in the world.

Ancistrocactus tobuschii is one of the fishhook cacti, so named because of some of the spines that are hooked. This species is very small, usually occurring only as single individuals. At most, the rather top-shaped, dark green plant is about three and one-half inches in diameter and has eight broad ridges, or ribs, each made up of pyramid-shaped swellings.

The area on a cactus plant from which the spines arise is called the areole. In Tobusch's fishhook cactus, there are usually three central spines produced at each areole. Two of the three central spines are straight, while the third one is hooked. Surrounding the central spines at each areole are seven to twelve straight and slightly shorter radial spines. The bright yellow flowers of Tobusch's fishhook cactus are most attractive. They are nearly two inches tall and broad.

Although this cactus, like all others, is subject to overcollecting, it has an additional threat to its existence because some of its populations grow near streams. A great flood in the area where some of the plants grow completely destroyed one of the populations of this cactus in 1978.

Nellie Cory Cactus or *Coryphantha minima.*
Cactus family. Succulent; stem to 1 inch tall and to ¾ inch wide; spines about 20 per areole, to ⅛ inch, pinkish turning to yellowish and finally to ashy-gray. Flowers rose, to ¾ inch across. Season: March.

What is the tiniest cactus in the world? The Nellie cory cactus (*Coryphantha minima*) of Brewster County, Texas, is a good candidate, but it doesn't get the blue ribbon. Nonetheless, this diminutive cactus is one of the smallest, as well as one of the rarest, cacti in the world.

Nellie cory cactus was first found by A. R. Davis in March 1931 on a ranch in the Trans-Pecos region of Texas, a few miles south of Marathon and several miles north of Big Bend National Park. Davis, who was a commercial dealer of cacti, had found it growing in the crevices of limestone at an elevation of about 4,000 feet. Amazingly, this species probably would not have been found except that it was in flower, for the tiny body of the cactus was growing under and completely concealed by the growth of a fern-related clubmoss called selaginella. Only the cactus's small rose-colored flowers penetrated above the selaginella.

Shortly after Davis's discovery, Ralph O. Baird, who was looking for cacti for the Garfield Park Conservatory collection in Chicago, paid Davis a visit in order to observe the plants in Davis's collection. Davis took Baird to see the tiny cactus growing in the limestone crevices, and Baird was convinced that it represented a new species, which he named *Coryphantha minima*.

For some reason, however, Davis had decided to send some specimens of his new little plant to Leon Croizat, a botanist at the Arnold Arboretum in Jamaica Plains, Massachusetts. Croizat, too, recognized the plant as a new species and named it *Coryphantha nellieae* in 1934, for Davis's wife, Nellie. Croizat apparently was unaware that Baird already had given this little plant a name three years earlier. The rules botanists observe governing the naming of plants dictate that the correct binomial for this plant is *Coryphantha minima*, even though the common name can still be Nellie cory cactus.

The Nellie cory cactus is usually less than one inch tall and only one-half to three-fourths inch wide. Each areole has up to twenty rather thick spines, which are about one-eighth inch long. The rose-colored flowers are only three-fourth inch tall and nearly as broad.

A second colony of *Coryphantha minima* was found on an adjacent ranch a few years after Davis's discovery. It occurred in an identical habitat. The owner of this ranch, however, permitted cactus collectors to enter his land in the 1960s and completely wipe out the plant.

Coryphantha minima was listed as a federally endangered species on November 7, 1979.

Bunched Cory Cactus or *Coryphantha ramillosa.*

Cactus family. Succulent; stems spherical, to 3½ inches in diameter; central spines 4, to 1½ inches, curved, gray mottled with brown; radial spines 14–20, curved and twisted, gray with dark tips. Season: April, May.

Nestled in the great curve of the Rio Grande of Texas is the rugged Big Bend country. Highlighted by the spectacular Chisos Mountains, the Big Bend is an area of wild mountains, rugged canyons, rocky deserts, and, of course, the great Rio Grande. A large part of the wilderness has been set aside as the Big Bend National Park, a park replete with scenery, history, and plant and animal life.

Many of the plants are unusual for the United States. Several of them are primarily Mexican, with the Big Bend area representing the northernmost extension of their range. More than sixty different kinds of cacti are known from Big Bend National Park, where they enjoy the protection provided by the National Park Service. Several of these cacti are very rare, and a few are on the federal list of endangered and threatened species.

One of the rarest of all the cacti is the bunched cory cactus (*Coryphantha ramillosa*), a small species confined to the Big Bend region. It was named in 1942 by Ladislaus Cutak. Cutak was a great authority on succulents for the many years he worked at the Missouri Botanical Garden in St. Louis. In his quest for succulents that could make good horticultural plants, Cutak visited the Big Bend country of Texas during the summer of 1938. Noting that the area was rich with cacti, he instructed several of the ranchers and local residents to be on the lookout for any plants that looked unusual, and to notify him if any such plants were found.

A year later, Mr. A. R. Davis, a cactus dealer from Marathon, Texas, sent Cutak specimens of a small cactus that Davis had found in the Big Bend area. This was the same Davis who later was to discover the Nellie cory cactus (*Coryphantha minima*). After studying the specimens, Cutak concluded that Davis's plants represented a new species, which he called *Coryphantha ramillosa,* the bunched cory cactus.

The plants that Davis sent to Cutak had been collected on the side and top of limestone hills at an elevation of about 3,500 feet. Since that original collection, the bunched cory cactus has been found in several remote canyons in the Big Bend area. Although most of the known locations for this species are on the United States side of the Rio Grande, this cactus has been found in north-western Coahuila, Mexico.

Cutak noted that the epithet *ramillosa,* meaning "with many branchlets," is well earned for this cactus, "for its armament greatly resembles a small bundle of dried twigs." Del Weniger, in his *Cacti of the Southwest,* described the spines as "untidy, curving and spreading at all angles, and making it look much like a bunch of dead, gray grass stems, which must be its particular form of camouflage."

The bunched cory cactus is a spherical plant with a diameter up to three and one-half inches. Each areole on the stem produces four central spines surrounded by fourteen to twenty radial spines. *Coryphantha ramillosa* forms gorgeous pink to deep rose flowers about two inches wide. The flowers are unable to open fully, however, because of the close crowding of the surrounding spines.

Because the once remote locations for this species are being threatened by development which could modify the habitat and make some specimens more accessible to collectors, the bunched cory cactus was officially placed on the federal list as a threatened species on November 6, 1979.

Lloyd's Hedgehog Cactus or *Echinocereus lloydii.*

Cactus family. Succulent; stems clumped together, cylindrical, to 1 foot long, to 4 inches thick, with 11–13 ribs; central spines 4–8, straight, purple-red to gray; radial spines 14–17, straight, purple-red to gray. Flowers lavender to scarlet, to 3 inches across. Season: February to April.

It is by a stroke of good fortune that Lloyd's hedgehog cactus (*Echinocereus lloydii*) has been able to avoid extinction.

This rather stout, beautiful-flowered cactus was first found by Mr. F. E. Lloyd late in February 1909 near Tuna Springs, Texas.

Bunched Cory Cactus
Coryphantha ramillosa

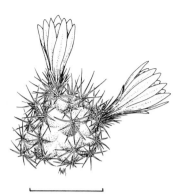

Davis' Green Pitaya
Echinocereus viridiflorus var. *davisii*

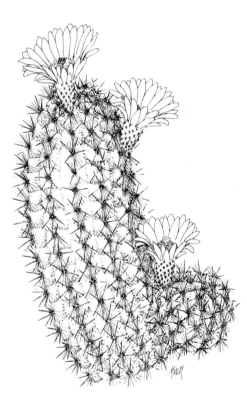

Lloyd's Hedgehog Cactus
Echinocereus lloydii

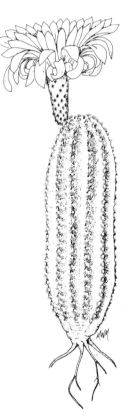

Albert's Black Lace Cactus
Echinocereus reichenbachii var. *albertii*

Lloyd collected the same species again in 1916, and sent plants to the New York Botanical Garden, where they were grown for several years.

All the while, Nathaniel Lord Britton and J. N. Rose were preparing their profusely illustrated four-volume classic, *The Cactaceae*, at the New York Botanical Garden. When they examined Lloyd's collections from Tuna Springs, they concluded that they were looking at a new species, which they named *Echinocereus lloydii* in 1922.

The good fortune alluded to earlier is that no one knew where Tuna Springs, Texas, was. Cactus dealers and rustlers, who usually flock to the locality of any newly named cactus, were beside themselves. Tuna Springs was not on most Texas highway maps. For nearly fifty years, Lloyd's hedgehog cactus grew undisturbed because of its secret location.

During the 1960s, when it was revealed that Tuna Springs was an old stagecoach stop in west Texas, cactus dealers rushed to the area to get their prizes. Then, when a highway-widening project wiped out another part of the population, the United States Fish and Wildlife Service took steps to give *Echinocereus lloydii* the federal protection it needed by naming it an endangered species on October 26, 1979.

Several stems of this cactus often grow clumped together. The largest stems may reach a height of one foot and may be as much as four inches thick. The rather large areoles have four to eight central spines and fourteen to seventeen radial spines surrounding them. All of the spines are straight. Few flowers match the brillance of those of Lloyd's hedgehog cactus. They are three inches across and range in color from lavender to scarlet.

Echinocereus lloydii grows in sandy or gravelly soils under very dry conditions at about 2,800 feet above sea level. Recently, this species has been found in southeastern New Mexico.

Albert's Black Lace Cactus or
Echinocereus reichenbachii var. *albertii.*

Cactus family. Succulent; stems cylindrical, elongate; central spine 1, to ¼ inch, black or dark purple; radial spines 14–20, less than ½ inch, white with dark purple tip. Flowers showy, rose-pink with dark red center, up to 3 inches across; petals jagged at tip, recurved backward after one day. Season: May.

One of the most popular kinds of cacti that is sought after and grown by cactus fanciers is the black lace cactus. The common name is derived from the lacy appearance of the spines. These cacti are further enhanced by their beautiful flowers.

The black lace cactus is known botanically as *Echinocereus reichenbachii*. It is a variable species, with several rather well defined variations known. In Texas, Lyman Benson, the noted American cactus authority, recognized six varieties. Each of these differs in characteristics of the spines. Most of the varieties of the black lace cactus in Texas grow in typically rocky habitats, primarily on either limestone or granite.

The variety known as Albert's black lace cactus (*Echinocereus reichenbachii* var. *albertii*) is remarkable because of its unique habitat requirements and its unusual spine characteristics. It grows under extremely heavy brush in the grassland areas southeast of Houston. The mesquite is so dense where the cactus grows that it forms impenetrable thickets. This actually has served as protection from overcollecting of this cactus.

In addition, since nearly all the known populations of Albert's black lace cactus occur on the large ranches common in this part of Texas, they have been spared the ravaging by cactus collectors, since the localities are on private property and therefore not accessible to the general public.

The chief distinguishing feature of Albert's black lace cactus is the single central spine of each areole, which is black or very dark purple and only about one-fourth inch long. It is surrounded by a ring of fourteen to twenty radial spines which lie close to the body of the plant. The flowers of this variety are extremely showy. They are rose-pink with a dark reddish center, and measure up to three inches across when fully open.

Although Albert's black lace cactus has been protected by virtue of its growing on private ranches, there is now a potential problem because the ranchers are using brush-killing chemicals and other means to clear the mesquite that has provided additional protection for the plant.

The common name is for R. O. Albert of Alice, Texas, one of the collectors of this variety.

Because of the limited number of populations known for Albert's black lace cactus, it was declared a federally endangered plant on October 26, 1979.

Davis' Green Pitaya or *Echinocereus viridiflorus* var. *davisii*.

Cactus Family. Succulent; stems solitary, nearly spherical, to 1 inch in diameter, with 6–9 ribs; central spines absent; radial spines 8–14, to ¼ inch, white or gray with reddish tips. Flowers pale yellow, to 1 inch long, never fully opening. Season: March.

It is coincidental that two dwarf cacti in two different genera grow together in the same habitat under identical conditions on the same ranches in Brewster County, Texas. Additionally, A. R. Davis of Marathon, Texas, was the first person to discover each of these plants. You already have met one of them, the Nellie cory cactus (*Coryphantha minima*), earlier in this chapter. The other is classified as *Echinocereus viridiflorus* var. *davisii*, and is commonly known as Davis' green pitaya.

Davis found this little cactus in about 1930 and showed it to A. D. Houghton, who named it a new species the following year. Houghton called it *Echinocereus davisii*, acknowledging that it was related to another species, *Echinocereus viridiflorus*, but differed sharply because of its diminutive size and the lesser number of ribs on the stem.

Nonetheless, W. T. Marshall, who made a study of *Echinocereus* during the 1940s, believed that Davis' green pitaya should be considered as a variety of *Echinocereus viridiflorus*, a decision accepted by the United States Fish and Wildlife Service.

Davis' green pitaya grows under mats of selaginella, the little clubmoss, in the same limestone crevices where the Nellie cory cactus grows. Like the Nellie cory cactus, this one also is unexposed until it pushes its yellow flowers up through the selaginella. One wonders how many other plants lay hidden, waiting only for a chance discovery by someone who happens to be in the right place at the right time. For example, there are two different kinds of Australian orchids that live their entire life, flowers, fruit, and all, completely underground. They were discovered when unearthed during farming operations.

Echinocereus viridiflorus var. *davisii* is nearly spherical and only about one inch tall and broad. There are no central spines; instead, each areole produces eight to fourteen radial spines not more than one-fourth inch long. The pale yellow flowers are only about one inch long and do not spread out into fully opened blossoms.

One of the populations known for Davis' green pitaya has been decimated by cactus collectors. Weniger, in his *Cacti of the Southwest,* noted that after its original locality was published in 1931, "collectors and dealers went straight to the spot and brought out hundreds of specimens. This continued for years. . . ."

Since the cactus is now confined to a single ranch in Texas, it was placed on the federal list as an endangered plant on November 7, 1979.

Lloyd's Mariposa Cactus or *Neolloydia mariposensis.*

Cactus family. Succulent; stem solitary, to 3½ inches tall, to 2 inches broad, blue-green to yellow-green, with 13–21 ribs; central spines 4–7, to ½ inch, bluish tipped; radial spines 25 or more, less than ½ inch, white or gray tipped with light brown. Flowers white or pink, to 1½ inches across. Season: May, June.

J. Pinckney Hester was one of the leading explorers of the Big Bend area of Texas for many years during the first half of this century. He was also an avid cactus enthusiast, who called his home in Fredonia, Arizona, "Cactus Haven."

In 1945 Hester was exploring a limestone ridge about twenty-five miles from Terlingua, Texas. The ridge overlooked an abandoned quicksilver mine. Hester discovered a little golf-ball-sized cactus growing in a thin layer of soil on hot, exposed limestone. He described his new species as *Echinomastus mariposensis,* a name changed later by Lyman Benson to *Neolloydia mariposensis.*

Lloyd's mariposa cactus grows singly. Its stems may reach a height of three and one half inches and a width of two inches, but are usually smaller. Each areole on the stem has four to seven central spines and twenty-five or more radial spines. The flowers, which are about one and one-half inches across, are sometimes white, sometimes pink. The problem that this and other cacti have had from cactus dealers is illustrated by the following account given by Weniger in his *Cacti of the Southwest.*

> It was one of the most terrible experiences of this study to come upon at least a thousand specimens of small cacti, mostly this rare species, gathered by someone and left to die in a pile on a hill only a few miles from the old . . . mine. I was told that this was probably a cache left by a professional cactus-digging crew for the

dealer to pick up with a truck—but a cache which he missed. At
any rate, the cacti were mostly burned to a crisp by the time I saw
them. . . .

With this sort of rapid extermination in mind, the United States
Fish and Wildlife Service designated Lloyd's mariposa cactus a
federally threatened species on November 6, 1979.

McKittrick Pennyroyal or *Hedeoma apiculatum.*
*Mint family. Matted perennial; stems to 6 inches, minutely hairy.
Leaves opposite, simple, very narrow. Flowers 1–3 in a cluster at
base of some leaves, pink, longer than the leaves. Season: July,
August.*

The mint family is made up of a well-defined group of plants that
occurs in many regions of the world, although it is not well represented in the Arctic. Most mints are readily recognized by their
opposite leaves borne along a four-sided, or square, stem. The
flowers are usually two-lipped, that is, two of the five united petals
form an upper lip which arches over the lower three petals. Several
species in the family produce oils which have a characteristic smell
and taste. Because of this, many mints, such as rosemary and
lavender, are grown in herb gardens.

Some of the most strongly scented mints are known as pennyroyals, and belong to the genus *Hedeoma*. Most of the more than
forty species which make up the genus live in the southwestern
United States and Mexico. One of these is the McKittrick pennyroyal (*Hedeoma apiculatum*).

Julian A. Steyermark, of *Flora of Missouri* fame, who was one
of the original discoverers of this species while he was still a
graduate student, was exploring the rugged canyons in the Guadalupe Mountains near the Texas-New Mexico border during August
1931. In crevices along a stream in McKittrick Canyon, Steyermark
found specimens of a small mint which was later to be named
Hedeoma apiculatum.

This species, which soon acquired the common name of the
McKittrick pennyroyal, is one of several unusual species that
botanists have found in McKittrick Canyon. The canyon is noted
for its scenery and its scientific importance. Along the canyon floor,
a stream appears and disappears several times. In the more shaded
parts of the canyon, ferns grow in abundance beneath a canopy of

Lloyd's Mariposa Cactus
Neolloydia mariposensis

Pipewort
Eriocaulon kornickianum

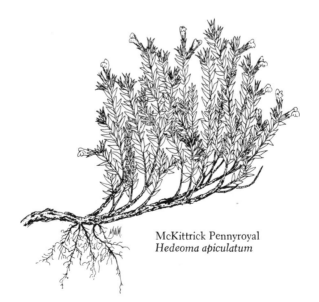

McKittrick Pennyroyal
Hedeoma apiculatum

big-tooth maple, walnut, hop hornbeam, and choke cherry. The massive north wall of McKittrick Canyon rises to an impressive height of 1,900 feet.

Because of its significance, McKittrick Canyon and adjacent areas were designated as Guadalupe Mountains National Park in 1965.

Most of the McKittrick pennyroyals known today occur in the national park, although there is one population on nearby private property and another group of plants across the state line in New Mexico's Lincoln National Forest.

This plant grows on limestone rocks and along streams at elevations between 5,000 and 8,000 feet. The biggest threat to its existence is the effect that hikers will have on it. Trails pass near many of the plants in McKittrick Canyon, and hikers who leave the trails to climb over the rocks may dislodge some of them.

The limited populations of this species prompted the United States Fish and Wildlife Service to list it as threatened on July 13, 1982.

The McKittrick pennyroyal stands only six inches tall and usually grows in mats. At the base of some of the uppermost leaves are clusters of one to three pink flowers.

Pipewort or *Eriocaulon kornickianum.*
Pipewort family. Perennial; stems solitary or in tufts. Leaves many, crowded near base of plant, to 2 inches, very narrow. Flowers crowded together in a single head on a 2- to 6-inch leafless stalk, the head gray and up to ¼ inch in diameter; sepals 2; petals 2, yellow. Season: July.

We were on our way to the mountains of Arkansas and the highest elevation between the Appalachians and the Rocky Mountains. These mountains are a haven for rare and unusual plants. We wanted to see the maple-leaved oak, the Rocky Mountain woodsia fern, and anything else that was unusual for this part of the country.

The route chosen for this foray from southern Illinois took us over the Mississippi River at Cape Girardeau and through a part of the scenic Ozark Mountains of Missouri. At Springfield, we turned south on U.S. Highway 65 and headed into Arkansas.

South of Harrison, we made several stops in the northern unit of the Ozark National Forest. At Alum Cove Natural Bridge, we stopped long enough to admire the 130-foot span of the shelter bluff and to observe the rich flora of the ravines. From Alum Cove, we drove through the heart of the wild Boston Mountains, emerging at Clarkville.

After circling the eastern half of the Dardanelle Reservoir, we proceeded to a rugged, massive mountain. The forest service road was steep in places as we climbed to the 2,650-foot summit. The mountain still shows signs of prior disturbance. The earliest settlers from the southern Appalachians lived off the land, and agricultural and timber practices removed much of the vegetation from the mountain's plateaulike summit. But despite these intrusions into the natural vegetation, most of the area along the margins of the plateau surface still support good natural plant communities.

Dry woods interrupted by prairie openings dominate the south face of the mountain. Plants typical of the prairies to the west and north of Arkansas thrive in the full sunlight this exposure receives.

Across to the shaded north-facing side of the mountain, a prominent bluff line with spectacular 200-foot drops highlights the area. It was on the north slope that I saw the maple-leaved oak, considered by botanists to be a variant of the Shumard red oak. This is the only location in the world for this unusual type.

Along the western side of the mountain summit is an area of small, spring-fed bogs. Sphagnum moss is the dominant plant of these mountaintop bogs.

Most of the very rare plants that we wanted to see were growing on the rugged rocky terrain. The Rocky Mountain woodsia fern was there, many miles from any other locality for this species. So, too, was the hay-scented fern (*Dennstaedtia punctilobula*), the prickly gooseberry (*Ribes cynosbati*), a thick-leaved sedum (*Sedum ternatum*), and an inconspicuous whitlow-wort (*Paronychia virginica* var. *scoparia*). The crinkled hairgrass (*Deschampsia flexuosa*) was growing on the summit. All of these plants are extremely rare in Arkansas.

Best of all, however, was a dwarf pipewort (*Eriocaulon kornickianum*), which was growing on the moist sand of an intermittent stream which passed through a bog. There are about 400 kinds of pipeworts, most of them growing in the warmer parts of the world. Several occur in the southeastern United States, but one hardly expected to find a pipewort on a mountain summit in Arkansas. In addition to this high mountain site, this little species occurs at two or three other localities in Arkansas, one in Oklahoma, and probably no more than three in Texas.

Eriocaulon kornickianum is a perennial growing as a solitary individual or in small tufts. The numerous leaves are crowded near the base of the plant. Several minute yellow flowers are crowded together in a solitary, almost spherical head which is mounted at the tip of a slender two- to six-inch-long, leafless stem. They bloom in July.

The United States Fish and Wildlife Service is reviewing this species for possible federal listing.

❋ 6 ❋

Dwindling Plants
of the Rocky Mountain States

The diverse habitats in the Rocky Mountain states, from high mountains to vast deserts, provide areas where rare species have found a niche in which they may survive. Twenty-one species from the Rocky Mountain states have already made the endangered or threatened species list; fourteen of them are cacti. The Rocky Mountain states include Montana, Idaho, Wyoming, Colorado, Utah, New Mexico, and Arizona.

Clay Phacelia or *Phacelia argillacea*.
Waterleaf family. Annual; stems to 1 foot, finely hairy. Leaves alternate, simple, deeply divided. Flowers bluish violet, in a curved cluster, bell-shaped, to ¼ inch. Season: July, August. (See color plate 23.)

Marcus E. Jones was a dominant figure in the botany of the Great Basin of Utah and adjacent areas, beginning in 1880. Jones, who was born in Ohio in 1852 and who had made a brief plant-collecting trip to Colorado in 1878, settled at Salt Lake City and began to make extensive collections of plants from the Great Basin. Traveling by train, bicycle, horse and buggy, and on foot, Jones explored remote areas previously unknown to botanists. As a result of these explorations, a great number of species new to science were discovered. From his own print shop, Jones personally hand-set the type for many of his publications. He founded his own private journal, entitled *Contributions to Western Botany*, in order to record his observations and to name his new species.

Sometime during August 1883 Jones arrived at Pleasant Valley

Junction in the Wasatch Mountains of Utah. Among his collections from that trip was a rather small, bluish-violet-flowered herb which Jones identified as the glandular phacelia, a plant found here and there in the western states. Jones was unaware that what he had found would later prove to be different enough from the glandular phacelia to be called a new species.

Eleven years later and some eleven miles to the east of Pleasant Valley Junction, Jones found another population of this new plant. Again he called it *Phacelia glandulosa*, the glandular phacelia.

Jones's plants from these two locations went unnoticed until 1971, when N. Duane Atwood of Brigham Young University, while working on some phacelias of the western United States, ran across the specimens Jones had collected and began to realize that they were not nearly as hairy and had smaller flowers and fruits than the glandular phacelia. Atwood sought to rediscover the phacelia by trying to retrace Jones's earlier expeditions. For several miles, Atwood walked along the Denver and Rio Grande Western right-of-way near what formerly was Pleasant Valley Junction. Success was met when Atwood discovered about fifteen plants growing on a gravelly hillside at an elevation of between 6,500 and 7,000 feet.

After studying his new collection and comparing it with Jones's earlier ones, Atwood concluded in 1973 that the plants represented a new species, calling it *Phacelia argillacea*.

The new species grows in a fine-textured clay derived from a shaley parent material known as Green River shale. This material is known for its abundance of well-preserved fossils of plants, insects, fish, and mammals. Since *Phacelia argillacea* grows in a shaley clay soil, it has acquired the common name of clay phacelia.

The clay phacelia is an annual whose finely hairy stems grow to a height of about one foot. Several bluish-violet flowers are arranged in a strongly curved terminal cluster. When Utah botanists revisited the clay phacelia site in 1979, only seven plants could be found. Before the end of the summer, three of these had been trampled to death by a flock of sheep, and two others were damaged. The few that remained were chewed on by rock squirrels later in the season.

A new search the next summer resulted in the discovery of about 200 plants across the highway and about one-fourth mile to the southwest of the original locality.

In addition to the threats posed by sheep and rock squirrels, there is concern that construction activities by the railroad could modify the habitat of the clay phacelia.

The clay phacelia was listed as a federally endangered species on September 28, 1978.

Rydberg's Milk Vetch or *Astragalus perianus*.

Pea family. Perennial; stems spreading or lying flat. Leaves alternate, compound, divided into as many as 19 oval leaflets to ¼ inch. Flowers 2–6 in clusters opposite the leaves, sweet-pea-shaped, white with pink tinge, to ⅓ inch. Pods broadly elliptic, inflated, purple-speckled, to 1 inch. Season: June, July.

Per Axel Rydberg came to the United States from Sweden in 1882 when he was twenty-two years old. He worked in the Michigan iron mines until an injury to his leg forced him to seek a less physically demanding occupation. He elected to become a teacher, and from 1884 to 1893 he taught at the Luther Academy in Wahoo, Nebraska. During his stint as a teacher, Rydberg became interested in plant exploration and made his initial collections from Nebraska in 1890. His major botanical projects were west of the Mississippi and resulted in the publication of two books which have served as a foundation for later studies. *Flora of the Rocky Mountains and Adjacent Plains*, published in 1917, and *Flora of the Prairies and Plains of Central North America* are landmarks in the botany of the western United States.

It was 1905, while searching for plants in conjunction with his Rocky Mountains flora, that Rydberg, accompanied by E. C. Carlton, visited Utah. In the vicinity of Marysvale, in what is now a part of the Fishlake National Forest, Rydberg and Carlton discovered a low-growing species of *Astragalus*, or milk vetch, a member of the pea family. The original collection was made not far from a remarkable yellow cliff on the west side of the Sevier River. The cliff later became popularized by the song "Big Rock Candy Mountain."

It wasn't until 1964 that the milk vetch Rydberg and Carlton had found was declared a new species. Rupert C. Barneby of the New York Botanical Garden noted that the plant had several features that were unlike those of any other known kinds of *Astragalus*. Barneby called this new species *Astragalus perianus*; at the time, it was known only from the original collection.

When the Smithsonian Institution made a survey of rare and vanishing plants of the Pacific Basin during the 1970s, they regarded Rydberg's milk vetch as highly endangered and probably extinct. But during the summer of 1975 this rare species was rediscovered.

It is known today from high-elevation habitats ranging from rocky clay soil to mountain woodlands or barrens to alpine meadows, between 10,000 and 11,500 feet above sea level.

Although *Astragalus perianus* seems fairly safe from immediate extinction, threats from mineral exploration and road construction loom in the background. There is considerable sheep grazing in areas where this species occurs.

Rydberg's milk vetch is a beautiful little perennial that grows from a thickened rootstock. Short clusters of two to six flowers arise opposite some of the leaves. The fruits are inflated pods that are shaped like tiny footballs. They are purple-speckled and a half inch to one inch long.

This plant was listed as a federally threatened species on April 26, 1978.

Dwarf Bear Poppy or *Arctomecon humilis*.

Perennial; stems to 3 inches. Leaves basal as well as alternate on the stems, simple, hairy, with three large teeth at the upper end. Flowers white, showy, to 1 inch. Season: May, June.

The poppy family is noted for the large, attractive flowers exhibited by most of its species. Ranking among the prettiest in the family is a small group of species from the western United States known as bear poppies, belonging to the genus *Arctomecon*.

There are only three species of bear poppies in the world, and all three are extremely rare. The first one ever found was the bright-yellow-flowered bear poppy. The collector was John C. Fremont, an explorer for the Corps of Topographical Engineers. On one expedition, while returning from California, Fremont passed through southern Nevada, where he discovered the bear poppy, which he sent to John Torrey. Torrey named it *Arctomecon californica*, for that part of Nevada was still considered California territory at the time. Ironically, *Arctomecon californica* has never been found in California, but only in Nevada and Utah. It is currently being reviewed by the United States Fish and Wildlife Service for possible endangered status.

A second species of bear poppy is *Arctomecon merriamii,* named by Frederick V. Coville, a noted botanist with the United States Department of Agriculture from 1888 to the early 1930s. He was botanist for the Death Valley Expedition in 1891. Coville named the poppy for the well-known geographer C. Hart Merriam, who had collected it in 1891 from a ranch in what is now the Las Vegas area. Merriam's bear poppy has large white flowers and is known only from Nevada and California. It, too, is being reviewed by the Fish and Wildlife Service for possible future listing.

This brings us to the third species, *Arctomecon humilis,* the dwarf bear poppy. Despite its much smaller size, it is equally beautiful. Charles C. Parry found this one in the vicinity of St. George, Utah, in 1874. Parry's career was an interesting one. Born in England, Parry came to the United States and eventually obtained a medical degree in 1846 from Columbia College. After practicing medicine for two years in Davenport, Iowa, Parry succumbed to his avocational interests in geology and botany. In 1848 he joined the geological survey of Iowa, Minnesota, and Wisconsin, and one year later he was named botanist for the Mexican Boundary Survey. For the rest of his life Parry spent his summers on various expeditions. In 1874, while attempting to follow John C. Fremont's earlier trail across southern Utah, Parry discovered the dwarf bear poppy.

This species has been found in a few locations in the area around St. George, Utah. During the 1950s, Peebles and Ruben found it just across the Utah border into Mohaje County, Arizona.

The dwarf bear poppy grows in very fine clay soil which dries out during the summer but which becomes a sticky gumbo during spring and autumn. The elevation is between 2,300 and 3,000 feet.

Because of this species' restricted habitat, it probably never has been a common plant, but adverse circumstances have led to a further decline. Several plants stood in the way of the building of the new city of Bloomington on the Virginia River south of St. George. In addition, the entire range for this species is in an area heavily used by off-road vehicles. Other potential threats include transplanting to gardens, mining of nearby gypsum deposits, and construction of highways and power lines. As a result,

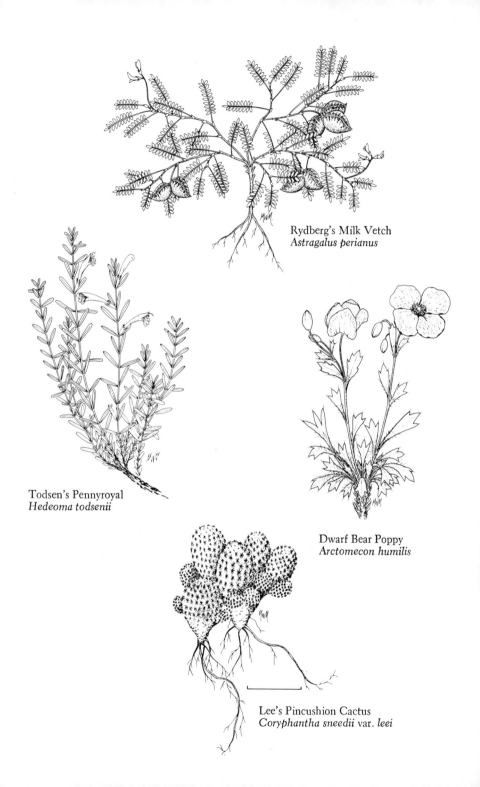

Rydberg's Milk Vetch
Astragalus perianus

Todsen's Pennyroyal
Hedeoma todsenii

Dwarf Bear Poppy
Arctomecon humilis

Lee's Pincushion Cactus
Coryphantha sneedii var. *leei*

the dwarf bear poppy was given federal endangered status on November 6, 1979.

Arctomecon humilis is a bushy little perennial which grows from a thickened base. Several inch-long white flowers that bloom in May and June are persistent for a long time on the plant.

Todsen's Pennyroyal or *Hedeoma todsenii*.
Mint family. Perennial; stems to 6 inches. Leaves opposite, simple, to ½ inch, very narrow. Flowers solitary in the leaf axils, red, to 1½ inches long. Season: August.

Many kinds of plants are restricted to growing on a particular type of rock or soil. Plants confined to a single kind of substrate which has a limited occurrence have, themselves, a very limited range. Thus it is with plants found exclusively on gypsum.

Chemically, gypsum is a soft hydrous calcium sulfate. It is found in various forms, from shiny white rocks to elegant crystals known as selenite to sparkling white grains of sandlike particles.

A visit to New Mexico's White Sands National Monument and adjacent areas will provide all the gypsum one could wish for. Since a few of the plants which have been listed by the United States Fish and Wildlife Service dwell only on gypsum soils, I wished to familiarize myself with different aspects of gypsum and its formation.

We pulled into the White Sands National Monument near Alamogordo, New Mexico, and stopped at the visitor's center, a must for anyone wishing to understand the many fascinating features of the monument.

All of the White Sands National Monument lies within a great depression known as the Tularosa Basin. The entire basin stretches for nearly 150 miles in a general north-south direction and is framed on three sides by rugged mountain ranges.

Looking directly west from the visitor's center, we could barely make out through the blowing sand the jagged outline of the San Andres Mountains. Through the centuries, infrequent rains have washed gypsum elements down the sides of the San Andres Mountains and onto the floor of the basin, the gypsum accumulating in an ancient geological lake. As the lake began to dry up because of climatic changes, these deposits of gypsum became exposed by the scouring action of severe winds. The great dune fields seen today at the White Sands National Monument have been built from this old lake's gypsum deposits, from gypsum in the groundwater which seeps up to the surface, and from new deposits continually washing down the San Andres Mountains after heavy rains and snows.

On the steep gypsum slopes of the San Andres Mountains is one of the rarest flowering plants in the country. It is a low-growing member of the mint family, called *Hedeoma todsenii*, or Todsen's pennyroyal. This plant was only recently discovered by Mr. T. K. Todsen on August 18, 1978. Todsen found it on a steep, north-facing gypsum hillside, growing in a habitat dominated by pinyon. There are estimated to be about 750 individuals of Todsen's penny-royal in existence. Since they occur on the White Sands Missile Range, they are afforded good protection, because the area is not accessible to most people. Future activities at the missile range in the vicinity of the plants, however, could possibly have adverse effects upon them. The Department of the Army has expressed a willingness to help protect this species. Nonetheless, the United States Fish and Wildlife Service listed it as a federally endangered species on January 19, 1981.

Hedeoma todsenii is a neat little perennial standing only about six inches high. The plant has showy red flowers which bloom in August.

Gypsum Wild Buckwheat or *Eriogonum gypsophilum.*

Buckwheat family. Perennial. Leaves basal, thick, nearly round, to 1 inch. Flowers several in a terminal cluster, small, yellow. Season: June to August. (See color plate 24.)

Another plant which is apparently able to grow only in gypsum soils is the gypsum wild buckwheat, *Eriogonum gypsophilum*. This species was found as early as 1909 in southeastern New Mexico by E. O. Wooton, one of the authors, along with Paul C. Standley, of the early *Flora of New Mexico*. The label on Wooton's speci-men reads "collected on a hill, growing in pure gypsum, August 6, 1909." Wooton and Standley described their new species as a perennial that grew from a thick, woody base. The neatly branched terminal cluster of bright yellow flowers was said to be up to six inches long.

For nearly three-quarters of a century, the gypsum wild buck-wheat has lived in this area of New Mexico, and nowhere else. It has survived grazing and off-road vehicles and botanists. A proposed lake for a water and power project in the plant's vicinity is on the drawing board, although preliminary estimates indicate that this rare species would be at least ten feet above the level of the lake. (Erosion alone could tumble some of the plants to their watery death.)

I was in the southwestern United States in early June and had hoped to see the gypsum wild buckwheat for myself. The literature and the original specimen indicated a flowering time of August, so I knew that if I was lucky enough to find the plant, I would have to settle for seeing and photographing only clusters of leaves.

Information I obtained from New Mexico State University at Las Cruces directed me to proceed from Artesia until a limestone hill came into view. Wooton and Standley had indicated that the hill was capped by fifty to one hundred feet of limestone, with the gypsum appearing in several layers in the lower two-thirds.

The road from Artesia goes through pretty flat terrain, and after several miles I was beginning to wonder if my information was correct. Suddenly a low hill appeared on the horizon ahead. When we got parallel to the hill, I pulled the van over to the shoulder and my sons and I climbed the embankment to cross the barbed-wire fence and head for the hill a few hundred yards away. Even before crossing the fence, Trent picked up a gypsum rock, and my hopes soared. We didn't make it to the hill—we didn't have to. Mark, who was in the lead, shouted, "I've found it!" while I was still trying to figure out how to get the other half of my body through the barbed wire. Finally disengaged from the barbs, I hurried to where Mark was bending over. There, growing among gypsum rock, were about ten separate tufts of the gypsum wild buckwheat. Better yet, they were all in full bloom, despite the fact that it was only June 14. The bright yellow blossoms literally glistened in the sunlight.

The rarity of the gypsum wild buckwheat and the potential threats to it have resulted in its listing as a federally threatened species by the United States Fish and Wildlife Service on January 19, 1981.

Sneed's Pincushion Cactus or *Coryphantha sneedii* var. *sneedii*.
Cactus family. Succulent; stems to 2 inches long and to 1 inch thick, clumped into colonies; central spines 13–17, to ⅓ inch,

white to pinkish; radial spines 24–50, to 1/6 inch, white to pinkish. Flowers pink to pale rose, to ½ inch tall, not opening fully. Season: May, June. (See color plate 25.)

It was a hot wind that caused the dust devils to boil up and move across the rolling hills above Las Cruces, New Mexico. The craggy peaks of the formidable Organ Mountains loomed in the distance. Our goal for the day was to track down a tiny, federally endangered cactus known as *Coryphantha sneedii* var. *sneedii*, the Sneed's pincushion cactus.

We hoped to get clues to the whereabouts of var. *sneedii* at New Mexico State University in Las Cruces. We asked for directions to the biology department and were happily surprised to find the herbarium just a few paces from the biology building's main entrance, instead of in the most remote corner of the uppermost floor, or in the basement, as it so often is. The three botanists who were working in the herbarium were eager to assist us in our mission to find var. *sneedii*. They told us to head for a mountain range outside of town.

We were directed to a limestone ridge with about a thirty-five-degree slope, in an area where the most conspicuous vegetation would be two kinds of desert spoons (*Dasylirion wheeleri* and *D. leiophyllum*), the tough-leaved sumac (*Rhus choriophylla*), and creosote bush (*Larrea tridentata*).

A limestone slope soon came into view, and we drove a rough gravel road to its end, then headed on foot across a pair of dry, rocky washes for about one quarter of a mile. Creosote bush was everywhere, and both species of the desert spoon were observed. Several different cacti were growing among the loose rocks on the slope, including another species of pincushion. Small colonies of a desiccated fern, brown but still erect, sprouted up among the white limestone rocks.

Perhaps fifteen minutes from the car, and forty feet upslope from the easternmost wash, the first little clump of rounded gray var. *sneedii* was observed. Once our eyes had become adjusted to this small plant, we were able to find a few more clumps here and there, and as we ascended toward the rocky crest of the slope, the plants became more common.

Sneed's pincushion cactus is a tiny white-stemmed cactus up to two inches long and less than one inch thick. The many white to pinkish spines that cover the stems give the plant a distinct whitish or pinkish appearance. At each areole there are thirteen to seventeen central spines and twenty-four to nearly fifty radial spines surrounding them. The pink to pale rose flowers, which bloom during May and June, are only about one-half inch high and do not open fully.

The first collections of this cactus were made by J. R. Sneed and sent to Nathaniel Lord Britton and J. N. Rose in 1921. Britton and Rose, working out of the New York Botanical Garden, were completing their monumental four-volume work *The Cactaceae*. They received the specimens in time to describe and name them in their encyclopedic work.

Upon publication of this new species, cactus collectors and dealers rushed to the mountains and quickly depleted the population. Some cactus dealers today sow this species from seed and offer it for sale, so that it is not necessary for people wishing to have their own Sneed's pincushion cactus to dig it up from the wild. This species also occurs in one other mountain range in New Mexico.

Sneed's pincushion cactus was declared a federally endangered plant on November 7, 1979, because of its very restricted range and because of excessive collecting by cactus merchants. It is known from just a few slopes in New Mexico. Most of the sites are on public property administered by the Bureau of Land Management. Since the plant is probably unable to expand its range, it is imperative to protect other specimens that are left from cactus poachers.

Lee's Pincushion Cactus or *Coryphantha sneedii* var. *leei*.

Cactus family. Succulent; stems to 2 inches long and to 1 inch thick, clumped into colonies; central spines 6–7, to ⅛ inch, white; radial spines many, minute, gray or white. Flowers pink, to ½ inch across, not opening fully. Season: May, June.

Lee's pincushion cactus is a larger relative of Sneed's pincushion cactus. It, too, is a very rare plant. It was discovered growing in one of the canyons of Carlsbad National Park, New Mexico, by W. T. Lee in 1924. Lee sent his specimens to the United States National Herbarium in Washington, where they remained unidentified until 1933, when they were finally recognized as a new kind of cactus.

During the succeeding years this plant has been found in a few other canyons, all in the Carlsbad area. Even though all the known populations occur in the national park, Lee's pincushion cactus is still in danger from illegal collecting. As a result, it was listed as a federally threatened plant on October 25, 1979.

Although the major attraction at Carlsbad National Park is the fabulous underground cavern, the above-ground habitats give a good picture of the natural regions of southeastern New Mexico. The seven-mile main park drive climbs from the entrance of the park to the escarpment where visitors enter the cave, passing through a variety of desert habitats. Elevations range from 3,600 feet above sea level in some of the canyons to 6,350 feet on top of Guadalupe Ridge. The canyon floors support vegetation dominated by black walnut, hackberry, various oaks, and the desert willow. Along the slopes and on the ridge tops a drier habitat prevails. Here, agaves, yuccas, sotols, ocotillo, and various kinds of cacti dominate the landscape. Huge colonies of the prickly pear cactus occur. Lee's pincushion cactus grows on limestone ledges of north-facing slopes.

For the first few years of its life Lee's pincushion cactus has a spherical stem, but as it matures, the stem elongates and becomes more or less club-shaped. At each areole on the stem is a cluster of six or seven white central spines. Surrounding these centrals are a great number of minute radial spines. The flowers, which fail to open widely, are about one-half inch broad. They bloom during the spring.

Nichol's Devil's Head Cactus or
Echinocactus horizonthalonius var. nicholii.

Cactus family. Succulent; stems solitary, nearly spherical, to 1 foot tall, blue-green, with 7–13 ribs; central spines 3, to 1 inch, one of them black and curving downward, the other two reddish and curving upward; radial spines 5, gray, curving, to ¾ inch. Flowers pink, to 2 inches across. Season: May. (See color plate 26.)

The genus of cacti called *Echinocactus* includes a famous group of plants known as barrel cacti. These take their name from the cylindrical, barrel-shaped stems of some of the species. Many stories abound about weary travelers, desperate for water, reviving themselves from the liquid stored in the stems of the barrel cactus.

There are other cacti besides barrels which belong to the genus *Echinocactus*. One of them is a fascinating plant called the devil's head, or turk's head, known botanically as *Echinocactus horizonthalonius*. There are several variations of the devil's head in the southwestern United States. The most uncommon of these is var. *nicholii*, or Nichol's devil's head cactus of southern Arizona. It is named for A. A. Nichol, who studied this plant while employed at the University of Arizona and then at the Arizona State Game and Fish Department. As early as 1930 Nichol suggested that this plant was different from the other devil's heads he had seen in Texas, but it wasn't until 1969 that Dr. Lyman Benson, working on *The Cacti of Arizona*, described it for the first time.

We wanted to relocate this unusual cactus, so we stopped by Arizona State University to see if we could get some clues as to where to look for it. Dr. Donald Pinkava gave us some information, and told us of several other rare Arizona plants he was working with.

Information we obtained indicated that our search for Nichol's devil's head cactus should center around a small range of mountains north of Tucson. The habitat where var. *nicholii* grew was north-facing limestone slopes where ocotillos, saguaros, and palo verdes occurred in abundance.

As we approached a likely area, saguaro and palo verde were noted as being abundant, but the ocotillo had not yet appeared, so we continued until we could see ocotillo growing about halfway up a north-facing slope with white limestone outcrops. Convinced that this might be the area for var. *nicholii*, we stopped the van along a dry wash and proceeded in the direction of the mountain a half mile away. The temperature already had soared above one hundred degrees on this midday in the middle of June. It was a steady climb to get to the limestone rocks, and we stopped a number of times, at my command, so that I could photograph some element of the desert. The steeper the slope became, the more often I found it necessary to take a photograph!

At last we reached the lowest white rocks on the slope. Amazingly, and true to form, the first Nichol's devil's head appeared, half buried in the dry, rocky soil. It was a beautiful plant, with the vertical rows of spines set against a blue-green stem.

From our position halfway up the mountain, we paused to view our surroundings. Despite the heat and dryness, we admired the beauty of the area. The mixture of ocotillos, saguaros, and other desert plants has always fascinated me, and the area we found ourselves in was equal to the best. In addition, the thrill of finding the cactus we had set out for with only a vague idea of where to go was very satisfying. We lingered on the slope for half an hour, finding a number of Nichol's devil's heads of various sizes. The bleached skull of a coyote, lying beneath a young saguaro stem, completed this truly western scene.

The largest specimen we saw was about ten inches tall and eight inches broad. Each areole has three central spines and five gently curving, gray radial spines. Nichol's devil's head produces attractive pink flowers about two inches across at the tip of the stem during the spring.

This cactus is known only from two locations in southern Arizona. It would appear that the major threat to it is probably over-collecting. There are considerable mining operations in the area, but not in the immediate vicinity of the cactus. Nonetheless, the restricted range of *Echinocactus horizonthalonius* var. *nicholii* justified its listing as a federally endangered plant on October 26, 1979.

Engelmann's Purple Hedgehog Cactus or *Echinocereus engelmannii* var. *purpureus.*

Cactus family. Succulent; stems usually several together, cylindrical, to 10 inches tall, to 2 inches across; central spines 4, purple, straight, to 1 inch; radial spines 6–12, straight. Flowers magenta or purple, to 2½ inches across. Season: February to May.

Anyone involved with the study of cacti soon becomes acquainted with the name of George Engelmann. Several cacti were named by Engelmann, and others, such as Engelmann's prickly pear and Engelmann's hedgehog cactus, bear his name.

George Engelmann was born in Germany but sailed for the United States in 1832 at the age of twenty-three, eventually settling in the area around St. Louis, Missouri, where he pursued a highly successful medical career. Engelmann was intensely interested in plants and soon became acquainted with Asa Gray, the great eastern botanist. Since St. Louis was the starting point for many exploring expeditions to the West during the middle 1800s, Engelmann was able to tell the plant collectors on these expeditions what to look for. In return, Engelmann had access to all these collections from the new frontier. Engelmann himself made several excursions to the West.

When the business entrepreneur Henry Shaw decided to establish a botanical garden in Mid-America, he sought the assistance of Engelmann, with the result that the Missouri Botanical Garden (known locally as Shaw's Garden) at St. Louis is one of the finest botanical institutions in the world today.

The genus *Echinocereus* is generally known collectively as the hedgehog cactus because of the extremely spiny nature of most members of the group. One of the common species of hedgehogs in the southwestern United States is Engelmann's hedgehog, *Echinocereus engelmannii.* Most cactus authorities are able to recognize several distinct varieties of Engelmann's hedgehog. Perhaps the rarest and most restricted one is known as Engelmann's purple hedgehog cactus, recognized by Dr. Lyman Benson as *Echinocereus engelmannii* var. *purpureus* in 1969.

This variety was discovered in 1949 by Benson, who found it in the vicinity of St. George in a part of the Mojave Desert in southwestern Utah. It was growing in sandy and gravelly soil derived from red Navajo sandstone. This cactus has not been seen since its discovery.

The Mojave is one of four major deserts in the southwestern United States generally recognized by geographers. It is the smallest of the deserts, occurring in the southern tip of Nevada and adjacent areas of southern California, northwestern Arizona, and southwestern Utah. The dominant plants throughout most of the Mojave Desert are the creosote bush (*Larrea tridentata*) and the bur sage (*Franseria dumosa*). A characteristic plant is the Joshua tree (*Yucca brevifolia*), a unique species which barely makes its way into the southwestern corner of Utah.

Engelmann's purple hedgehog cactus grows in the desert shrub community of the Mojave Desert at an elevation of about 2,900 feet.

This cactus usually branches near the base to form as many as ten cylindrical stems that grow up to eight inches tall and up to two inches in diameter. There are usually four central spines and six to twelve radial spines. The magenta or purple flowers are about two and one-half inches wide.

Engelmann's purple hedgehog cactus is widely sought by cactus growers. On October 11, 1979, it was listed as a federally endangered plant.

Kuenzler's Hedgehog Cactus or *Echinocereus kuenzleri*.

Cactus family. Succulent; stems solitary or in small clusters, to 6 inches long, to 4 inches across, with 10 ribs; central spine 1, twisted, to 1 inch, with a whitish covering; radial spines 4–5, curved, to ½ inch, with a whitish covering. Flowers purplish pink, to 4 inches long. Season: May.

The Sacramento Mountains are an imposing range that rises quickly to the east of Alamagordo, New Mexico. Alamagordo lies in the Tularosa Basin, a flat, dry region with near-desert conditions. It is a centering area for trips southward into White Sands National Monument and the White Sands Missile Range. To the east, U. S. Highway 82 climbs out of the basin and into cool, coniferous forests in the Sacramento Mountains as it makes its way to the resort community of Cloudcroft in just twenty miles.

It was at the eastern edge of the Sacramento Mountains where Horst Kuenzler found a colony of about twenty hedgehog cacti in 1961 that didn't match up with any of the known hedgehogs. Some of the original colony was dug up and grown for several years in a private garden in Belen, New Mexico. A specimen from this garden was sent to the University of New Mexico, where Drs. Edward Castetter, Prince Pierce, and Karl Schwerin finally named it *Echinocereus kuenzleri* in 1976. In the meantime, Kuenzler and Dale Morrical of Las Cruces found a few additional colonies in the general vicinity of the original collection. Ironically, the colony that was first found in 1961 was completely destroyed a few years ago during road improvements.

The stems of Kuenzler's hedgehog may occur singly or in groups of as many as eight. The usual size for this cactus is about six inches long and four inches wide. The single central spine at each areole is about one inch long and often twisted. Surrounding it are four or five curved and sometimes twisted radial spines about one-half inch long. During May, the tip of each stem produces a three- or four-inch-long purplish-pink flower.

The United States Fish and Wildlife Service estimates that fewer than 200 plants of Kuenzler's hedgehog still live in the wild.

With this greatly reduced number, the plant was declared a federally endangered species on October 26, 1979.

Arizona Red Claret Cactus or
Echinocereus triglochidiatus var. *arizonicus*.

Cactus family. Succulent; stems several in a cluster, cylindrical, to 9 inches tall; central spines 1–3, dark gray; radial spines 5–11, curved, pinkish tan. Flowers bright red, to 2 inches across. Season: May. (See color plate 27.)

The morning dawned crisp and clear as we prepared to break camp near Show Low, Arizona. The tops of the tall pines waved back and forth in the gentle breeze. While my wife was preparing breakfast, my sons and I took a brisk hike along the Mogollon Rim, sharpening our eyes for plants in anticipation of an exciting day ahead.

By midday we hoped to be in the Globe-Superior area, searching for some rare species that grow in the stony mountains of that famous copper country.

To reach Globe, we traveled beautiful Highway 60 through the spectacular Salt River Canyon. This is unlike the Arizona familiar to many. Instead of desert, the terrain is mountainous and the vegetation dominated by pleasant-scented pines.

The Mogollon Rim separates the high country of northern Arizona from the low country to the south. The road down the Mogollon Rim south of Show Low is not as abrupt as some places where the rim is crossed. Highway signs warn of a five-mile downgrade as Salt River Canyon is approached. Each turn in the road brought a new spectacular scene before us. We stopped at the pullout across the river and looked up at the towering cliffs and then across to the Salt River. How amazing that such a relatively small river today could have caused this 2,000-foot gorge.

The motor of our van droned as it pulled our camper up the five miles on the other side. The white slag of an abandoned asbestos mine shone in the distance.

We drove through the mining towns of Globe and Miami, everywhere surrounded by the sterile mountains of mining slag. A few miles east of Superior we entered the scenic Queens Canyon, located in the Tonto National Forest. The road through the

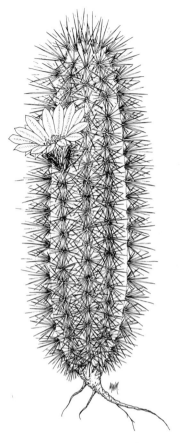

Kuenzler's Hedgehog Cactus
Echinocereus kuenzleri

Engelmann's Purple Hedgehog Cactus
Echinocereus engelmannii var. *purpureus*

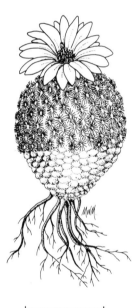

Spineless Red Claret Cactus
Echinocereus triglochidiatus var. *inermis*

Brady Cactus
Pediocactus bradyi

gorge culminates in a modern tunnel, followed by a picturesque bridge over the Queen River.

Beyond Superior, Picketpost Mountain, a craggy monolith, is outlined against the blue sky. The Boyce Thompson Southwestern Arboretum is nestled at the base of the mountain.

The arboretum was founded in 1929 and is operated by the University of Arizona, the Arizona State Parks Board, and the Boyce Thompson Arboretum, Inc. Col. William Boyce Thompson was a copper-mining magnate whose dream was to create a living museum for plants that could grow in the high Sonoran Desert country. He began his private arboretum at his winter home at the foot of Picketpost Mountain in the 1920s. At the formal dedication of the arboretum in 1929, Prof. Franklin Crider noted that the arboretum's purpose was "to grow every tree, shrub, vine, grass, and flower able to withstand in the open ground the climate of the Southwest; . . . a place where the yet hidden secrets of these varied forms may be further revealed for the enrichment of human knowledge and the benefit of mankind." The arboretum today is a welcome oasis in the Sonoran Desert.

At the arboretum, we interrupted Kent Newman's lunch hour. Kent is an employee of the Arizona State Parks Board and has been at Boyce Thompson more than a decade. His specialty is cacti, and he had volunteered to show us the federally endangered Arizona red claret cactus, *Echinocereus triglochidiatus* var. *arizonicus*.

Kent climbed into the van and directed us to the foothills of the Pinal Mountains. We climbed above the tilted limestone, past palisadelike rhyolite columns, and into an area of granite boulders. Kent stopped us where a rock jutted out to the road, and we made our way up a steep, gravelly rocky passage, crawling over the smaller boulders and around the larger ones. The prickly manzanita (*Arctostaphylos pungens*) stabbed us from the left, and the leaves of the holly-leaf buckthorn (*Rhamnus californica*) scraped us as we tried to slip by them. For good measure, *Mimosa biuncifera*, known as the wait-a-minute plant, clawed at our clothing, causing us to stop and disengage our pants legs from it.

Kent and my sons were about fifty yards upslope from me when Kent called out that he had found a nice clump of *Echinocereus triglochidiatus* var. *arizonicus*. I hastened to the spot and beheld a marvelous group of ten stems growing in front of a large granitic boulder. I took several photos of the handsome cactus while the rest of my party searched for a colony which might still have a lingering flower or two. The search was in vain, for this variety had bloomed about the last of May, nearly two weeks before our visit.

The habitat was exceptionally dry and rocky. A few colonies of the Arizona red claret were found, but the plant was definitely uncommon. The United States Fish and Wildlife Service has estimated that about a thousand individuals are all that remain, and they are constantly being sought by collectors. This variety was listed as federally endangered on October 25, 1979.

Variety *arizonicus* usually grows in several-stemmed colonies, with the largest stems attaining a height of nine inches and a diameter of about six inches. At each areole are one to three dark gray central spines and five to eleven pinkish-tan radial spines. Bright red flowers about two inches broad make this cactus too conspicuous when it flowers during May.

Spineless Red Claret Cactus or
Echinocereus triglochidiatus var. *inermis.*
Cactus family. Succulent; stems several in a cluster, to 4 inches tall, to 2 inches thick, with 8–9 ribs; spines usually absent. Flowers bright red, to 2 inches across. Season: May.

The red claret cactus (*Echinocereus triglochidiatus*) is a highly variable cactus distributed through much of the southwestern United States. The most unusual-looking variety is called var. *inermis,* and it is so strange-looking that, at first glance, it does not seem to be a red claret cactus. *Inermis* means "unarmed," referring to the fact that the stems have few or no spines. The plants look naked and unprotected in their severe habitat.

Known as the spineless red claret cactus, this variety has been found at a few locations in west-central and southwestern Colorado and Utah at an elevation between 5,000 and 8,000 feet. It is usually found at the base of pinyon pines, where it grows in the acidic debris dropped by the pines. It grows to its fullest expression when in partial shade.

This spineless variety was first discovered during the nineteenth century, but was seen so infrequently after that that there were suggestions it might be extinct.

During his work on the cactus flora of Colorado, Gerald Arp made an attempt to relocate living populations of *Echinocereus triglochidiatus* var. *inermis* and met with considerable success. He found during his observations that the spineless variety often grows in the vicinity of other spiny varieties of *Echinocereus triglochidiatus*, and even tends to intergrade with them. It is true that some specimens of var. *inermis* do have a few very short spines.

Each colony of this spineless cactus is composed of five to fifteen stems which are only about four inches tall and about half as wide. Although this variety may not have gotten its full share of spines, it did get its complement of beautiful flowers. The bright red blossoms, which open during May, are about two inches across when fully open.

The unusual spineless nature of this variety makes it highly sought as a collector's item. Since the cactus is known to grow in areas rich in mineral and oil deposits, it also could face danger in the future from mineral and oil exploration and extraction. This variety was listed as federally endangered on November 7, 1979.

Brady Cactus or *Pediocactus bradyi*.
Cactus family. Succulent; stem solitary, to 3 inches long, to 2 inches broad, mostly underground; central spines none; radial spines numerous, to ¼ inch, white or yellow. Flowers pale yellow, less than 1 inch across. Season: March.

No group of cacti can be any more fascinating than the cluster of eight tiny species that comprise the genus *Pediocactus*. In general, all of the species of *Pediocactus* are very small, some only about the diameter of a nickel. They frequently just protrude above the surface of the desert, and during extremely dry periods they may retreat completely into the ground, making them virtually impossible to find. All eight species grow in the vicinity of the Four Corners region where Utah, Colorado, Arizona, and New Mexico come together.

The recent detailed work by Ken Heil, Barry Armstrong, and David Schleser of Navajo Community College at Shiprock, New Mexico, has done much to unravel the problems in the genus. Their work not only included a study of the plants themselves, but also the habitats, geology, soil types, and elevation.

Of the eight species of *Pediocactus*, four have already been listed as federally endangered species, and three others are currently under review for possible listing. Only Simpson's cactus, *Pediocactus simpsonii*, is common enough to escape listing.

The Colorado River has been responsible for creating some of the most beautiful scenery in the world. From the unequaled beauty of the Grand Canyon to the highly popular and scenic Glen Canyon Recreation Area in northern Arizona, the Colorado River flows past and through other scenic marvels such as the colorful Vermilion Cliffs and the picturesque Marble Canyon. It is in this general area that the Brady cactus (*Pediocactus bradyi*) resides. Just looking for this little species is a thrill in itself because of the marvelous surroundings. The strikingly beautiful Vermilion Cliffs are always in the background as we stop to admire various cacti which grow in the area.

During the heat of summer, the diminutive Brady cactus pulls down into the soil until the stems are scarcely distinguishable from the limestone pebbles which are scattered on the desert floor. The stems of *Pediocactus bradyi* are usually solitary and less than three inches long and two inches broad. They are densely covered with white or yellow radial spines.

This species remains virtually concealed until the summer rains arrive in August and September. The plants then expand and later form their flower buds, which will finally open in March of the following year. The flowers are pale yellow and less than an inch long and broad.

Although this species occurs sparingly in the Four Corners area, it does not grow anywhere else in the world. Highway construction and power-line maintenance have destroyed some plants, but overcollecting is the most important threat to this species. It is likely that some specimens were wiped out when the Glen Canyon Dam was erected. Brady cactus was listed as a federally endangered species on October 26, 1979.

Knowlton's Cactus or *Pediocactus knowltonii.*

Cactus family. Succulent; stems solitary or in dense clusters, to 2 inches long, mostly underground; central spines none; radial spines 18–26. Flowers pink, to 1 inch across. Season: March.

The smallest and probably the rarest *Pediocactus* is the Knowlton cactus, *Pediocactus knowltonii*. This inch-wide plant occurs only

1 / Small Whorled Pogonia
Isotria medeoloides

2 / Spreading Globe Flower
Trollius laxus

3 / Mountain Golden Heather
Hudsonia montana

4 / Bunched Arrowhead
Sagittaria fasciculata

5 / Persistent Trillium
Trillium persistens

6 / Chapman's Rhododendron
Rhododendron chapmanii

7 / Tennessee Coneflower
Echinacea tennesseensis

8 / Green Pitcher Plant
Sarracenia oreophila

9 / Canebrake Pitcher Plant
Sarracenia alabamensis

10 / Price's Groundnut
Apios priceana

11 / Oconee Bells
  Shortia galacifolia

12 / Southern Yellow Orchid
  Platanthera integra

13 / Leafy Purple Prairie Clover
  Petalostemum foliosum

14 / Northern Monkshood
  Aconitum noveboracense

15 / Lake Iris
  Iris lacustris

16 / Great Lakes Thistle
Cirsium pitcheri

17 / Houghton's Goldenrod
Solidago houghtonii

18 / French's Shooting Star
Dodecatheon frenchii

19 / Synandra
Synandra hispidula

20 / Mead's Milkweed
Asclepias meadii

21 / Kankakee Mallow
Iliamna remota

23 / Clay Phacelia
Phacelia argillacea

22 / Texas Poppy Mallow
Callirhoe scabriuscula

24 / Gypsum Wild Buckwheat
Eriogonum gypsophilum

25 / Sneed's Pincushion Cactus
Coryphantha sneedii var. *sneedii*

26 / Nichol's Devil's Head Cactus
Echinocactus horizonthalonius var. *nicholii*

27 / Arizona Red Claret Cactus
Echinocereus triglochidiatus var. *arizonicus*

28 / Peebles Navajo Cactus
Pediocactus peeblesianus

29 / Debris Milk Vetch
Astragalus detritalis

30 / Graham's Beardstongue
Penstemon grahamii

31 / Rollins' Thelypody
Glaucocarpum suffrutescens

32 / Lemon Lily
Lilium parryi

33 / Golden-Chested Beehive Cactus
Coryphantha recurvata

34 / Red Lobelia
Lobelia laxiflora var. *angustifolia*

35 / Yellow-Spined Barrel Cactus
Ferocactus acanthodes var. *eastwoodiae*

36 / Eureka Dunes Evening Primrose
Oenothera avita ssp. *eurekensis*

37 / Raven's Manzanita
Arctostaphylos hookeri ssp. *ravenii*

38 / Salt Marsh Bird's Beak
Cordylanthus maritimus ssp.
maritimus

39 / Cobra Plant
Darlingtonia californica

40 / Hawaiian Wild Broad Bean
Vicia menziesii

along the Los Pinos River a few miles east of the Four Corners Monument. It has been found in both Colorado and New Mexico. The elevation in this area is about 6,500 feet. The Knowlton cactus grows in flattish terrain dominated by pinyon pine, juniper, and sagebrush.

In his publication resulting from a study of *Pediocactus* in 1961, Dr. Lyman Benson relates the story of the discovery of this species.

This cactus was sent to the writer in the spring of 1958 by the late Mr. Fred G. Knowlton of Bayfield, Colorado. In his notes, Mr. Knowlton indicated that "there were acres of coarse gravel that 'dozer had ramped around in. Found a few dead plants but 'dozer has cut everything down and little white spined balls were coming up all over the place. . . . Am sending a couple—Didn't know there were any Pedio within miles."

Dr. Benson dedicated this species to Mr. Knowlton's memory after Knowlton died in a fire which destroyed his home only a month after his finding of the cactus.

Following the original discovery, Ken Heil of Navajo Community College and Prince Pierce of the University of New Mexico have found the Knowlton cactus a few more times, but all in the same general location. Since the habitat in which it grows extends down as far as Navajo Lake, it is surmised that the impoundment of the lake probably destroyed additional localities for this species.

The Knowlton cactus may occur singly or in small, dense clusters. The slightly elongated stems may reach two inches long, but most of the stem remains buried in the soil. Each areole has from eighteen to twenty-six radial spines. The pink flowers, which usually bloom in March, are up to one inch broad and long.

The greatest threat to this small species is from overcollecting. It was listed as a federally endangered species on October 26, 1979.

Peebles Navajo Cactus or *Pediocactus peeblesianus.*
Cactus family. Succulent; stems solitary or in small clusters, spherical, to 2½ inches in diameter, often mostly underground; central spines none; radial spines 3, less than ½ inch, curved, pale gray. Flowers yellow to yellow-green, to 1 inch across. Season: March. (See color plate 28.)

Another of the rare species of *Pediocactus* is Peebles Navajo cactus, *Pediocactus peeblesianus*. Peebles Navajo cactus was first found in 1935 by Mr. W. Whittaker of the Arizona Highway Department, who found it in the vicinity of the Arizona Agricultural Inspection Station near Holbrook.

In 1937, Colonel Bumstead and Herb Bool of Phoenix found this cactus in the same general area, but when personnel from the Desert Botanical Garden in Phoenix searched for it later, they found only several trampled dead plants. However, in 1947, J. Pinckney Hester discovered several plants of Peebles Navajo cactus on the other side of Holbrook. This cactus, which grows on low hills covered by quantities of round river gravel, lies obscured among the pebbles.

During the summer the plant retreats into the ground, with only the tip lying at the surface of the desert. Following late summer rains and the snows of winter, the plant expands and remains above the soil line until the flowers are produced in March.

The stems of Peebles Navajo cactus grow singly or in small clusters. There is no central spine at the areoles, but there usually are three white to pale gray, recurving radial spines. The nearly inch-wide flowers are yellow to yellow-green.

Like most species of *Pediocactus*, Peebles Navajo cactus has never been very common because of its extremely restricted range. The range has become even narrower due to the construction of Interstate Highway 40 and a coal-burning power plant. In addition, off-road vehicles and livestock have devastated a great expanse of habitat for this cactus. Gravel-pit operations and overcollecting for horticultural purposes also have added to the reduction of it. This species became listed as federally endangered on October 26, 1979.

Siler Pincushion Cactus or *Pediocactus sileri*.

Cactus family. Succulent; stems solitary or in small clusters, cylindrical, to 6 inches; central spines 3–7, straight, to 1 inch; radial spines 11–15, straight, to ¾ inch. Flowers yellow to yellow-green, to 1 inch across. Season: March to May.

Unlike the other endangered kinds of pediocacti discussed earlier in this chapter, *Pediocactus sileri*, known as the Siler pincushion cactus, is a much larger species that does not draw itself back into the soil during the arid summers. In fact, it resembles a very small barrel cactus because of its cylindrical shape.

An interesting fact about this species is that it occurs in two distinct areas (although within fifty miles of each other) and in two distinct soil types, and its spines have two distinct color phases.

A. L. Siler found the first specimens of this cactus in May 1883 at Cottonwood Springs and Pipe Springs in northern Arizona. The plants represented a new species, which was called *Echinocactus sileri.* (It wasn't until 1961 that Dr. Lyman Benson transferred the species to the genus *Pediocactus.*) For nearly one hundred years this species has existed in the same general area around Pipe Springs National Monument in Arizona to west of the Hurricane Cliffs in Utah.

In some places, *Pediocactus sileri* grows in white gypsum-containing soils; in other areas, it grows in soils with a reddish color derived from parent sandstone rock. Coincidentally, the plants which grow on gypsum soils tend to have white spines, while those that grow on the red, sandy soils have darker spines.

The stems of the Siler pincushion cactus occur singly or in small clusters. Spines cover the stem completely and often persist for many years, the older ones forming a sort of frayed and withered skirt around the base of the plant. Each areole has from three to seven nearly straight central spines up to one inch long and eleven to fifteen slightly shorter radial spines. Several yellow to yellow-green flowers about one inch wide appear near the tip of the stem during the spring.

Most of the Arizona population is relatively secure by virtue of its location in Pipe Springs National Monument or on the Kaibab Indian Reservation, where plant collecting is forbidden. The Utah plants, however, are more prone to overcollecting and off-road-vehicle trampling. In the future, gypsum-mining operations and the Warner Valley Power Project could further endanger this plant.

Biological problems that this cactus experiences include the fact that rabbits forage on it and burrowing rodents attack it from beneath, eating the plant tissues and leaving only a hollow shell above ground. As a result, the United States Fish and Wildlife

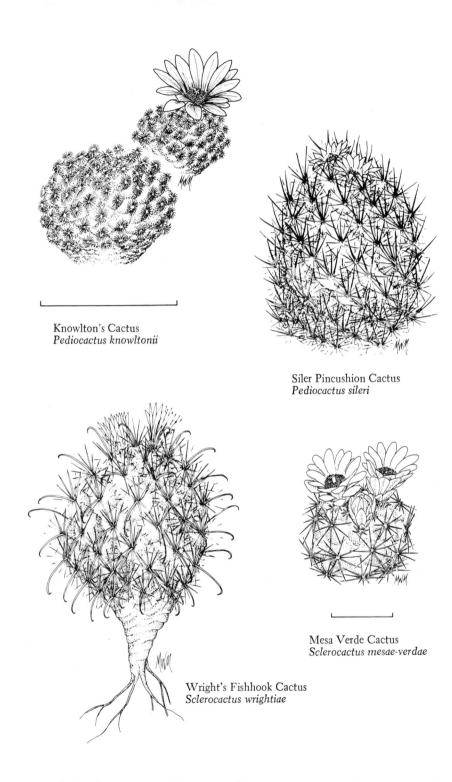

Knowlton's Cactus
Pediocactus knowltonii

Siler Pincushion Cactus
Pediocactus sileri

Wright's Fishhook Cactus
Sclerocactus wrightiae

Mesa Verde Cactus
Sclerocactus mesae-verdae

Service declared *Pediocactus sileri* to be federally endangered on October 26, 1979.

Wright's Fishhook Cactus or *Sclerocactus wrightiae*.

Cactus family. Succulent; stem spherical, to 3½ inches in diameter, with 13 ribs; central spines 4, the longest one to ½ inch, curved, with a hook at the tip; radial spines 8–10, straight, to ½ inch. Flowers mostly white, to ¾ inch across. Season: April, May.

Sclerocactus has been a very misunderstood genus because of the characteristics which it shares with other genera of cacti. Although Britton and Rose recognized *Sclerocactus* as a genus as early as 1922, it was not until 1966, when Dr. Lyman Benson wrote a definitive account of *Sclerocactus*, that the genus became accepted by most American botanists.

Probably the rarest of the species of *Sclerocactus* is *Sclerocactus wrightiae*, Wright's fishhook cactus. It occurs only in a handful of sites in the Navajo Desert of east-central Utah.

Mrs. Dorde Wright of Salt Lake City first found this species in 1961 in Emery County, Utah, where it was growing on dry, desert hills at an elevation close to 5,000 feet. Later, Dr. Irving G. Reimann of the University of Michigan found it several miles south in gravelly banks along the Fremont River.

When the federal government declared Wright's fishhook cactus as endangered on October 11, 1979, it indicated that this cactus was known only from five sites in two Utah counties. The species is being threatened by mineral exploration, potential industrial use of the habitat, off-road-vehicle activities, and overcollecting.

Sclerocactus wrightiae has a nearly spherical, unbranched stem up to three and one-half inches long. Each areole has four central spines and eight to ten white radial spines. During the spring, white flowers about three-fourths inch in diameter open from the top of the stem.

Mesa Verde Cactus or *Sclerocactus mesae-verdae*.

Cactus family. Succulent; stem spherical to elongated, to 3 inches, gray-green, with 13–17 ribs; central spines usually ab-

sent; *radial spines 8–11, pale, to ½ inch. Flowers yellow to greenish-yellow, to 1 inch tall, funnel-shaped. Season: May, June.*

Who cannot be impressed by the wonderfully preserved ruins that are found at Mesa Verde National Park in the extreme southwestern corner of Colorado? Standing in the midst of Spruce Tree House and looking out across the landscape, one can easily imagine the life-style of the Indians who lived in this region for 1,200 years.

Mesa Verde National Park is situated in the Navajoan Desert, where the arid conditions support a vegetation dominated by pinyon pine and juniper. Here and there are cacti of various sorts. Back around 1940, a cactus of a different sort was found in the Mesa Verde area by Charles Boissevain, who at the time was preparing his book *Colorado Cacti* with Carol Davidson. When Lyman Benson studied this plant, he called it *Sclerocactus mesae-verdae.*

For several years the Mesa Verde cactus was not found again. In recent years it has been found at a number of locations in the Four Corners region. Several of its sites are in the Navajo Indian Reservation.

Sclerocactus mesae-verdae is a gray-green cactus with spherical or slightly elongated stems about three inches tall. Central spines are usually lacking from each areole, but eight to eleven radial spines are present. The yellow to greenish flowers of the Mesa Verde cactus do not spread out when open; instead, they are almost funnel-shaped.

The United States Fish and Wildlife Service listed this species as federally threatened on October 30, 1979. The plants are subjected to overcollecting, as well as to threats from highway construction and off-road vehicles.

Uinta Hookless Cactus or *Sclerocactus glaucus.*
Cactus family. Succulent; stems solitary or in small clusters, spherical, to 2 inches in diameter, blue-green; central spines 1–3, straight; radial spines 6–8, straight. Flowers pink, to 2 inches across. Season: May.

In the general area from Grand Junction, Colorado, to Roosevelt, Utah, including a portion of the Uinta Basin, grows a very rare cactus known as the Uinta hookless cactus, *Sclerocactus glaucus.*

This cactus grows in dry, almost desert conditions in the gravelly soils of hills and mesas.

Sclerocactus glaucus has been found only a few times throughout the region of western Colorado and eastern Utah. Carl A. Purpus discovered this species in the Mesa Grande area of Colorado in June 1892. Marcus E. Jones was the first to find the Uinta hookless cactus in Utah while he was collecting on the south side of the Uinta Mountains near Roosevelt, in May 1908. In 1933, while doing a study of the plants of the Uinta Basin for the Carnegie Museum, Edward H. Graham found it again in the vicinity of the Green River.

Sclerocactus glaucus has single stems or clusters of two to nine stems which are blue-green and up to two inches in diameter. There are one to three straight, inch-long central spines per areole, surrounded by six to eight slightly shorter radial spines. Near the top of the stems during May are found pink flowers which measure up to two inches wide.

Sclerocactus glaucus was listed as federally threatened on October 11, 1979.

Handsome Phacelia or *Phacelia formosula*.

Waterleaf family. Annual; stems to 10 inches. Leaves alternate, simple, deeply divided, grayish. Flowers violet, in a coiled cluster, to ½ inch. Season: July, August.

Walden, Colorado, lying south of the Wyoming border by eighteen miles, is surrounded on all sides by unique and interesting land formations and natural habitats.

Less than twenty miles to the west is the rugged Mount Zirkel Wilderness, which straddles the Continental Divide in the Routt National Forest. Scenery is unmatched in the 72,000-acre wilderness, whose plant communities range from spruce-fir forests to alpine meadows. Fourteen peaks with elevations greater than 12,000 feet are blanketed with snow much of the year. There are only about thirty frost-free days over most of the area.

Fifteen miles to the east of Walden, in the Medicine Bow Mountains, is the Rawah Wilderness in the Roosevelt National Forest. Rawah, meaning "wilderness" in the Ute Indian language, is an area of high-elevation uplands with thin soil that supports fragile vegetation.

Several miles south of Walden, the well-known Rabbit Ears Peak has served as a landmark to travelers for nearly two centuries.

Northeast of Walden are the North and East Sand Dunes, perhaps the most unique topographic feature of all in the vicinity. The East Sand Dunes, a registered Colorado Natural Area, is one of only two active cold-climate dunes in Colorado. The area is in a fragile state of soil and vegetation development.

The Michigan River flows past Walden just a short distance from town. In places, sandstone bluffs line the river. It was along one of these sandstone bluffs, in loose sandy soil, that George Everett Osterhout made the original collection of the handsome phacelia, *Phacelia formosula*, on August 6, 1919.

The handsome phacelia is an annual whose stems generally branch from the base of the plant. Several pretty, half-inch-long violet flowers are arranged in a tight cluster on one side of the strongly coiled tips of the branches. The flowers bloom during late July and early August.

Although several botanists have observed this plant at its only known location since Osterhout's discovery—including Osterhout himself a second time on August 6, 1930—the handsome phacelia is in grave jeopardy of departing from our world. When a field survey was made on July 25, 1980, only 117 mature specimens and 3 seedlings could be found. They were in four different patches along about one-fourth mile of the sandstone bluff. The bluff receives heavy use by off-road motorcyclists, whose activities erode the fragile habitat.

As a result, the United States Fish and Wildlife Service proposed endangered status for *Phacelia formosula* on September 2, 1980. The proposal is pending at the time of this writing.

Heliotrope Milk Vetch or *Astragalus montii.*

Pea family. Perennial; stems to 2 inches. Leaves alternate, compound, divided into 5–13 elliptic leaflets to ⅓ inch. Flowers 2–8 in a cluster, sweet-pea-shaped, pinkish-purple with a white tip, to ⅓ inch. Pod egg-shaped, inflated, speckled, to ⅔ inch. Season: June, July.

One of the largest genera of flowering plants in the world is *Astragalus*, the milk vetches which belong to the pea family. Current estimates give the number of species of *Astragalus* at greater than 1,500. As might be expected with a group this enormous, there is considerable variation within the genus.

The milk vetches have evoked considerable interest among botanists. Marcus E. Jones of Utah published one of the first major works on the genus in 1923. His work, however, has been superseded by that of Rupert C. Barneby, of the New York Botanical Garden. Barneby, who has been studying the genus intensively for four decades, not only has brought order out of chaos in *Astragalus*, but he has collected extensively in the West and discovered many new species of plants himself.

Several species of *Astragalus* are rare and being considered by the federal government for listing. One of these is a recently discovered plant from the Wasatch Plateau of central Utah.

In 1978, Dr. Stanley L. Welsh of Brigham Young University named a new species of milk vetch he had discovered two years earlier. This species, which Dr. Welsh called *Astragalus montii*, was found growing on limestone in Sanpete County, Utah, at an elevation of about 11,000 feet.

This perennial grows from a thickened, woody, forked root. At the tip of each short stem is a cluster of two to eight pinkish-purple flowers. From each flower is produced a speckled, egg-shaped, inflated pod one-half to two-thirds inch long.

The heliotrope milk vetch is confined to about eighty acres in the Manti-LaSal National Forest. Since it seems to be unable to tolerate habitat disturbance, it might be susceptible to sheep which graze the area and off-road vehicles which have been seen on the summit of the mountain where it grows. Accordingly, the federal government proposed this species for listing as endangered on January 13, 1981.

Debris Milk Vetch or *Astragalus detritalis*.

Pea family. Perennial; stems very short or nearly absent. Leaves alternate, compound, divided into 3–7 narrow leaflets. Flowers several in a cluster, bright rose-purple, sweet-pea-shaped, to 1 inch. Pods flattened, to ½ inch. Season: May, June. (See color plate 29.)

The Uinta Basin forms a well-defined topographical unit in northeastern Utah and adjacent northwestern Colorado. It is a huge natural depression surrounded by a rim dominated by the crest of the Uinta Mountains to the north and the Book Cliffs to the south. Near the western edge of the basin is Strawberry Reservoir, while the eastern boundary is just across into Colorado in the Gray Hills. The ruggedly beautiful Green River bisects the area as it passes through Desolation Canyon.

The Uinta Basin is noted for its shale outcrops and sandstone formations. Most of the region is of the mixed desert and shrub type, although the dry, rocky, often desolate terrain caused Dr. Reed C. Rollins of Harvard University to write that the area "more nearly approaches a badland type of habitat."

Mineral resources are rich in the Uinta Basin, with large supplies of natural gas in the floor of the basin, great deposits of coal, and oil-rich shale. Because of this, mineral exploration is a continuous hazard for the plants that occur in the basin.

One of the showiest endemics in the Uinta Basin is a brilliant rose-colored member of the legume family known as the debris milk vetch (*Astragalus detritalis*).

Marcus E. Jones, the tireless western botanist who discovered many species in previously unexplored areas, first encountered this brightly flowered milk vetch in the vicinity of Roosevelt (at the time called Theodore), Utah, in 1910. Then, in 1933, Edward H. Graham, on an expedition of the Carnegie Museum of Pittsburgh, found it again in a shaley habitat dominated by a common shrub known as shadscale.

More recently, *Astragalus detritalis* has been found in several locations in the Uinta Basin in both Utah and Colorado. Once considered an endangered species by Utah botanists, it has been downgraded to threatened.

This species is a low-growing perennial which has woody, branching roots. There are almost no stems, the leaves arising nearly at ground level. The brilliant rose-purple sweet-pea-shaped flowers are borne several in a cluster during May and June.

Graham's Beardstongue or *Penstemon grahamii*.

Snapdragon family. Perennial; stems to 8 inches, hairy. Leaves opposite, simple, oval, toothless, to 2 inches. Flowers in terminal clusters, lavender with red-purple stripes, to 2 inches. Season: May. (See color plate 30.)

Even before the first white settlements in the form of stock ranches were made in the Uinta Basin in 1873, a few botanists had made some collections of plants from the area. Some plants had been collected on John Wesley Powell's expeditions down the Green and Colorado rivers in 1869 and 1871. The most extensive collections were probably made by Sereno Watson, who served as botanist on the 1869 Geological Exploration of the 40th Parallel.

Until 1931, however, there were only sporadic collections by botanists in the Uinta Basin. In that year, as well as in 1933 and 1935, Edward H. Graham and his wife were sent by the Carnegie Museum to collect the plants of the 12,000-square-mile Uinta Basin. One plant which Graham found is one of the most attractive species covered by this book. It is *Penstemon grahamii*, or Graham's beardstongue, of the snapdragon family.

Graham discovered this species in full flower in a talus slope on the west side of the Green River on May 27, 1933. He found it a little further north in both flower and fruit. Since Graham's discoveries, this species has been found a very few times in gravelly sandy soil over shale in the same area as the original discoveries.

Larry England of the United States Fish and Wildlife Service has told me that *Penstemon grahamii* grows in shale so rich in oil that oil virtually drips from the plant when it is removed from the soil. Industrial development and mineral exploitation of the habitat pose potential threats to this species, and it is being reviewed by the United States Fish and Wildlife Service. It is considered endangered by botanists.

There are more than 300 kinds of Penstemons, or beardstongues, in the world, most of them with rather large, pretty flowers. They are abundant in the western United States.

The marvelous thing about Graham's beardstongue is how many large flowers there are compared to the size of the plant. Each hairy stem grows only six to eight inches tall but bears nearly two-inch-long flowers in crowded terminal clusters. One has to smile when looking at a fully opened flower because of its fanciful resemblance to a face with a wide-open mouth, orange tongue, and lower row of teeth! The flower is basically lavender in color, but red-purple stripes, touches of white, and a little bright orange add vividness to each flower. There are several hairy, toothless leaves toward the base of the plant.

Rollins' Thelypody or *Glaucocarpum suffrutescens.*

Mustard family. Tufted perennial; stems to 8 inches, smooth. Leaves alternate, simple, elliptic, toothless. Flowers in elongated terminal clusters, yellow; sepals 4; petals 4, yellow, to ½ inch. Pods very narrow, to 1 inch. Season: May, June. (See color plate 31.)

How can a plant whose scientific name is *Glaucocarpum suffrutescens* obtain the unlikely common name of Rollins' thelypody? The story goes something like this.

Edward H. Graham and his wife were studying plants in the Uinta Basin of Utah for the Carnegie Museum in 1931, 1933, and 1935. On May 23 of his final year in the basin, Graham discovered a short, yellow-flowered plant on the eastern slope of a mountain about fifteen miles east of Desolation Canyon in Uintah County, Utah.

The plant was obviously a member of the mustard family because of its floral characteristics, but the genus and species could not be determined by use of any available flora. Graham sent the specimen to Dr. Reed C. Rollins of Harvard University, who was just beginning his illustrious career as the foremost authority on the mustard family in the United States.

Rollins quickly concluded that the plant was a new species, but the absence of fruits made it impossible to tell precisely what genus to place it in. Since the plant had a few similarities to *Thelypodium,* a western group of mustards, Rollins described the new species as *Thelypodium suffrutescens.* Even while describing it, Rollins noted that "the generic position of this species is not certainly determinable without mature fruiting specimens, although the nearest relative is probably *Thelypodium elegans. . . .*"

Since most species of *Thelypodium* have the common name of thelypody, and since Dr. Rollins gave the plant its Latin name, this species became known as Rollins' thelypody.

Rollins was not convinced about the true relationship of this species, and the only way he could discover its true identity would be by collecting it in a fruiting condition. On June 15, 1937, Dr. Rollins arrived in the Uinta Basin and headed for the mountain where Graham had first discovered the plant. By Rollins's own account, he "found the plants growing on a narrow (twenty feet) highly calcareous stratum of shale of the Green River formation. The plants . . . were traced along the face of a high bluff . . . for more than three miles."

Rollins was there at the right time, because the tufted plants were in fruit. What he found were plants with a slightly woody base, without basal leaves, with well-developed nectar glands, and with long, narrow fruits. Not one of these characteristics was a feature of *Thelypodium*. Rollins concluded that the plant was a member of a new genus, which he named *Glaucocarpum*, and the species became known as *Glaucocarpum suffrutescens*.

The habitat for this species is in the pinyon-juniper community in mixed desert shrub. For a while it was believed to be extinct, but recently it has been rediscovered in an area marked with mineral claims. Its location in a potential oil shale development area makes it extremely vulnerable. The United States Fish and Wildlife Service is reviewing it for possible federal listing as an endangered species.

The plant is a tufted perennial that grows from a deeply penetrating root. The smooth stems are about eight inches tall and bear several alternate leaves. Several yellow flowers are borne in a terminal, elongated cluster.

Colorado Eutrema or *Eutrema penlandii.*
Mustard family. Perennial; stems to 7 inches. Basal leaves ovate, heart-shaped at base, long-stalked; leaves on stem oblong, without leaf stalks. Flowers several per stem, to ⅛ inch; sepals 4; petals 4, white. Pods elliptic, to ⅓ inch. Season: July.

We were spending part of the summer of 1981 in the Pike National Forest west of Colorado Springs in glorious Rocky Mountain scenery. After botanizing the area around Pikes Peak on the eastern side of the forest, we moved our base of operations to the western section of the forest near the old western town of Fairplay.

One of the areas I was interested in getting to was Hoosier Ridge, a high-elevation natural area along the Continental Divide about eight miles south of Breckenridge. Hoosier Ridge is one of Colorado's Registered Natural Areas. A unique topographical feature of Hoosier Ridge is that there, the Continental Divide runs in an east-west direction for a short distance. The area is

Uinta Hookless Cactus
Sclerocactus glaucus

Handsome Phacelia
Phacelia formosula

Heliotrope Milk Vetch
Astragalus montii

Colorado Eutrema
Eutrema penlandii

well above timberline. One of the remarkable habitats is a moist bog covered with an abundance of mosses. From among the mosses grow a few kinds of flowering plants that are disjunct from arctic and other alpine areas far to the north.

Leaving Fairplay, we traveled a short distance to Alma and decided to take a short side trip to the Windy Ridge Ancient Bristlecone Pine Scenic Area. The road was rough and narrow as we climbed beyond 10,000 feet in elevation. A sharp curve brought us face to face with a gentle mountain slope dotted with the tortuous and gnarled bristlecone pines, the oldest living things in the world. The pines on Windy Ridge are widely spaced in a meadowlike setting. I tingle with the excitement of a new discovery every time I see a particularly ancient and picturesque specimen of bristlecone pine. The most spectacular occur high in the White Mountains of California's Inyo National Forest. Another striking group is found in the brilliant red rock formations between Navajo Lake and Strawberry Point in Utah's Dixie National Forest. On Windy Ridge, these pines have lived for hundreds of years in the crisp, cool mountain air of western Colorado.

We retraced our route to Alma, then headed north to Hoosier Pass. The pullout along the highway at the pass is at an elevation of above 11,500 feet. It was a climb of another 1,000 feet before we could locate any moss-covered bogs. After considerable searching, we found a small group of short, fruit-bearing herbs that obviously belonged to the mustard family. The plant was *Eutrema penlandii*, one of the high alpine species we were looking for.

Eutrema is a genus comprised of only two species. One of them is found around the northern polar area, where it occurs in North America from Greenland to Alaska. The other species is *Eutrema penlandii*, known only thus far from Park and Summit counties, Colorado. C. William T. Penland first found this species near Hoosier Pass, where it was growing in seepage below a snowbank, on July 27, 1935. Until 1967, the Hoosier Ridge area was the only known location for *Eutrema penlandii*. In 1967, William Weber of the University of Colorado discovered it on two adjacent mountain slopes. During the last five years, several new locations have been discovered, primarily by Barry C. Johnston of the United States Forest Service.

Eutrema penlandii is a smooth perennial growing only to a

height of seven inches. The leaves crowded at the base of the plant are ovate, heart-shaped at the base, and on long stalks; those on the stem are oblong and without leaf stalks. Each plant produces several white flowers only about one-eighth inch long.

Because of the recent collections of this species, *Eutrema penlandii* is considered a sensitive species by Colorado botanists.

Lemon Lily or *Lilium parryi*.

Lily family. Perennial; stems to 5 feet. Leaves alternate or whorled, simple, narrow, toothless. Flowers yellow, to 3½ inches long, 6-parted. Season: June, July. (See color plate 32.)

Dr. Charles Christopher Parry, an English-born physician who practiced medicine in Davenport, Iowa, spent much of his time as a botanist on various expeditions to the West. His first botanical assignment was to collect plants for the United States and Mexican Boundary Survey from 1849 to 1852. This brought him into contact with southern California, an area he revisited several times to collect plants and finally just to spend his winters in the latter years of his life.

In 1876, on perhaps his last botanical foray in southern California, Parry decided to honor a long-standing invitation to visit two brothers he had befriended earlier. The brothers, J. G. and F. M. Ring, operated an extensive potato farm in a mountain nook north of San Gorgonio Pass and east of San Bernardino. By Parry's own account, he wished to see the potato ranch, which was at an elevation of above 4,000 feet.

Leaving San Bernardino on horseback, Parry crossed Mill Creek and began to climb the steep ascent along the Yucaipa Ridge, finally arriving at an area which overlooked the broad sweep of San Gorgonio Pass. There were scattered groves of Coulter's pine (*Pinus coulteri*), and the gigantic nine-by-six-inch cones lying on the forest floor amazed Dr. Parry.

In one of the areas of moist, rich soil in the mountains, the Ring brothers had laid out their potato ranch. Both bachelors, the brothers spent much of their free time following scientific and cultural pursuits. When Parry arrived for his visit, he found that the home possessed an excellent library composed largely of scientific works and books on exploration and travel, and a battery of instruments for keeping meteorological data.

After resting his horse, Parry and his hosts began to explore the steep, gravelly slopes adjacent to the potato ranch. In one location near the house, Parry discovered a lily with lemon-yellow flowers. Unable to identify the lily, Parry eventually sent it to a colleague, Sereno Watson, who was studying the California flora. Watson named the plant in his friend's honor, calling it *Lilium parryi.*

Since its original discovery, the lemon lily has been found in four counties in southern California, where the California Native Plant Society considers it to be a rare species. It has also been found in the Santa Rita and Huachuca mountains in southern Arizona.

My first finding of the lemon lily capped a glorious week during the summer of 1979. My wife, three teen-age children, and I were about to fulfill a naturalist's dream by spending a week at the Mile-Hi Ranch, a retreat owned and operated by The Nature Conservancy south of Sierra Vista, Arizona. The ranch is located in Upper Ramsey Canyon along a babbling mountain brook, an oasis in the Huachuca Mountains in an otherwise arid, desert region.

Mile-Hi is a naturalist's paradise, where plants and animals live in a dense tropical setting only minutes from the adjacent desert. The chief attraction at the ranch is the variety of hummingbirds. Fourteen different kinds have been seen at the Mile-Hi, where dozens of feeders have been placed outside the rustic cabins to ensure a continuous food supply.

One nippy June morning we decided to hike the trail leading up Ramsey Canyon. Our guide was Jim Anderson, the ranch naturalist, who the night before had wowed us by showing beautiful wildlife slides he had taken at Mile-Hi.

All along the shaded stream that paralleled the trail were gorgeous wildflowers. In one exceptionally wet area, Jim called us over to see the lemon lily (*Lilium parryi*). It was a thrilling sight, and Jim told us that this lily was found at only one other place in Arizona.

The lemon lily stands as much as five feet tall and bears several long, narrow leaves. The lowest leaves on the stem are arranged in a whorled cluster, while the other leaves are alternately arranged. Two to ten pale yellow flowers are produced during June and early July at the tip of the stem.

Golden-Chested Beehive Cactus or *Coryphantha recurvata*.
Cactus family. Succulent; stems usually in clumps, cylindrical, to 10 inches; central spines 1–2, gray, curved downward; radial spines 12–20, slightly curved. Flowers greenish-yellow, to 1½ inches across. Season: May, June. (See color plate 33.)

It was a hot but clear day when we left Nogales, Arizona, on the Mexican border for Sycamore Canyon, a rugged wilderness I had read about that was twenty-seven road miles to the west. Sycamore Canyon is a precipitous ravine that extends for nearly five miles in a north-south direction until it reaches the Mexican boundary. The area is known for its many rare plants and animals, most of which are Mexican species that penetrate into the United States only as far as Sycamore Canyon.

Shortly after leaving Interstate Highway 19 a few miles north of Nogales, our blacktop highway turned into a narrow, rough, gravel road. We reluctantly passed several interesting-looking canyons that led off to the left. We finally arrived at a shady little oasis surrounding a spring called Hank and Yank Springs, where we parked under a spreading Emory oak. It was early afternoon and the temperature had reached well above one hundred degrees.

Leaving the van, we soon came to a few broken-down adobe walls, the remains of the Bartlett Ranch, a once prosperous cattle ranch that was the site of an Apache raid in 1886.

A small gravelly wash, dry during the summer, marked the beginning of Sycamore Canyon. A sharp bend in the wash brought us in view of hundred-foot pinnacles, an inviting entrance to the ravine that lay beyond. A wild cassava, or tapioca, plant (*Manihot angustifolia*), which I never dreamed might be in Arizona, was growing from the rocks. A couple of agaves, a shrubby honeysuckle, and various cacti and succulents related to sedum were observed.

About a quarter of a mile into the canyon from where we parked, the United States Forest Service has designated 545 acres which are exceptionally rich in plant and animal life as the Goodding Research Natural Area. This notable area is named for Leslie N. Goodding who made thorough studies of Sycamore Canyon during the 1930s and '40s. On a rocky ledge we discovered the golden-chested beehive cactus, *Coryphantha recurvata*. This cactus is known from a few places in Sonora, Mexico, and up into Sycamore Canyon. The plants we were looking at did not form a massive colony, but each individual stem was cylindrical and completely obscured by the spines. There were one or two gray central spines which curved downward very sharply from each areole. Around them were twelve to twenty slightly curved radial spines. Earlier in the year when the plant had flowered, there were greenish-yellow blossoms up to one and one-half inches wide.

Coryphantha recurvata is not on the federal list of endangered and threatened species, but its very restricted distribution makes it an excellent candidate for consideration.

Red Lobelia or *Lobelia laxiflora* var. *angustifolia*.

Lobelia family. Perennial; stems to 2 feet. Leaves alternate, simple, narrow. Flowers several, bright red, long-tubular. Season: May, June. (See color plate 34.)

We had found several interesting plants on our exploration into Sycamore Canyon west of Nogales, Arizona, and just above the Mexican boundary.

I was surprised, therefore, to run across a trailing member of the pea family which I was familiar with in the dry, rocky woods of southern Illinois. The plant was the attractive butterfly-pea (*Clitoria mariana*), whose two-inch-long lavender flowers are among the largest in its family. Several plants were sprawling over crumpled layers of rocks.

While most of the rare plants of Sycamore Canyon are found in the rocky cliffs along the gravel wash, there are interesting flowering plants that grow in and along the gravel of the creek. At one place where the creek drops from a two-foot-tall ledge, a plant of the tropical malpighia family was growing. I identified the specimen as a member of the genus *Aspicarpa*, but I have yet to determine what species it is, because it doesn't quite "fit" the Aspicarpas known from Arizona.

The showiest plant in the gravel wash, however, is the red lobelia (*Lobelia laxiflora* var. *angustifolia*). During late May, the cluster of red flowers literally glows.

This plant is very uncommon in Arizona, where it is found sparingly only in Sycamore Canyon. Leslie Goodding discovered it there in 1935. Across the border into Mexico and down into other countries of Central America, the red lobelia is found in oak-dominated woodlands. Since it has a broad range, even though it is rare in the United States, it has not yet been considered for review by the United States Fish and Wildlife Service as a possible endangered or threatened plant.

This lobelia is a perennial that stands about two feet tall and has several narrow leaves arranged alternately along the stem. Numerous bright red flowers are arranged in an elongated cluster. Each flower, which has an asymmetrical shape, is narrowed into a long slender tube which has been adapted for pollination by hummingbirds.

Yellow-Spined Barrel Cactus or
Ferocactus acanthodes var. *eastwoodiae.*
Cactus family. Succulent; stems cylindrical, to 3 feet or more, with 20 ribs; central spines 4, straight, to 3 inches, yellow; radial spines 12–14, to 2½ inches, yellow. Flowers yellow tinged with red, to 2½ inches across. Season: May. (See color plate 35.)

Fascinating among the cacti of the southwestern United States are those thick, cylindrical-stemmed ones called barrel cacti. Barrel cacti are grouped into two genera, *Ferocactus* and *Echinocactus*. One rather widespread species of *Ferocactus* which ranges from Arizona to California and south into Mexico is the common barrel cactus, *Ferocactus acanthodes*. Most plants of this species have red, gray, or even white spines, but one very rare variety has conspicuous yellow spines. It has been named var. *eastwoodiae*.

We had left the Boyce Thompson Arboretum near Superior, Arizona, with Kent Newman serving as our guide. Our primary goal was to find and photograph the Arizona red claret cactus (*Echinocereus triglochidiatus* var. *arizonicus*). A short distance from the arboretum, a wooden sign bearing the words "Apache Tears" pointed down a one-lane road that headed off across the dry, rocky terrain.

Apache tears are black, glasslike particles of basalt, volcanic in origin, which are found among deposits of perlite. The perlite deposit found in the Superior area is the largest in the country. It is mined and shipped to California, where it is expanded and sold for horticultural purposes as a soil lightener. The legend about the Apache tears is that when the Indians around Superior were being rounded up, the braves felt that it would be better to die as martyrs than to be captured by the white man. Many of them leaped to their death from a high cliff nearby, and the shiny Apache tears represent the tears shed as a result.

We passed up a chance this time to search for Apache tears, since we had a rare cactus on our mind. After a few minutes' drive, we pulled over where the highway shoulder was wide. Across a deep ravine rose a vertical, craggy cliff. Kent directed our gaze about halfway up the cliff. There, projecting from the rock ledges at precarious angles, were yellow barrel cacti! By using our binoculars, we could easily see the bright yellow spines covering the stems.

"That's the yellow-spined barrel cactus," remarked Kent. "Most cactus people call it *Ferocactus acanthodes* var. *eastwoodiae*."

Dr. Lyman Benson had named this variety in 1969, probably from plants growing near the same colony we were looking at this hot June day. Part of the amazing story about this plant is that it was growing within one hundred yards of a major Arizona highway, some plants nearly two feet tall, yet it hadn't been given a scientific name until 1969. One of the reasons for this, no doubt, was that this cactus grows in virtually inaccessible places, out of the reach of even botanists. Ken told me that it grew only in this area and again over in the Organ Pipe Cactus National Monument, some 175 miles to the southwest, as the cactus wren flies.

The yellow-spined barrel cactus may reach a few feet tall and up to one foot thick. The yellow spines are so dense that they practically conceal the stem. At each areole are four central spines up to three inches long and twelve to fourteen radial spines which are a little shorter. The flowers bloom regularly during May and irregularly after that. They are yellow tinted with a little red and up to two and one-half inches across.

Although most of the wild specimens of this cactus grow on hard-to-reach cliffs, cactus hunters will sooner or later find a way to get to them. As a result, the yellow-spined barrel cactus should be listed as an endangered plant.

Disappearing Wildflowers
of the Far Western States

Nineteen different kinds of plants from the far western states of Nevada, Washington, Oregon, and California have been listed as federally endangered or threatened species. Five of these are from California's Channel Islands. Included in this chapter are Macfarlane's four-o'clock and the Osgood Mountain milk vetch, which also occur in Idaho.

Eureka Dunes Evening Primrose or
***Oenothera avita* ssp. *eurekensis*.**
Evening primrose family. Perennial; stems to 2 feet, white-hairy. Leaves alternate, simple, triangular, white-hairy. Flowers solitary in the leaf axils, showy; sepals 4, pointing backwards; petals 4, white, to 1 inch, becoming reddish as they wither. Season: March to May. (See color plate 36.)

It was a memorable week that we spent on the eastern flank of the Sierra Nevadas exploring and observing the natural features of the Inyo National Forest. With Bishop as our headquarters, we made forays in every direction to see a multitude of contrasting habitats. One day we drove west out of Lone Pine as far as the road up the mountain could take us to Whitney Portal, the starting point to the lower forty-eight's highest mountain at 14,495 feet. On another occasion we drove past the resort community of Mammoth Lakes, through colorful mountain meadows below the jagged peaks of the Minarets, and down a narrow dirt road to the Devil's Postpile National Monument, an unbelievable geological formation of closely compacted, three- to seven-angled vertical columns.

A third venture led us northeast from Bishop to see the world's greatest and oldest stand of bristlecone pines in the White Mountains. The road was good as it climbed quickly to about 10,000 feet to the first group of bristlecones, known as the Schulman Grove. A nature trail maintained by the United States Forest Service passes Pine Alpha and other pines that are probably more than 4,000 years old. Several of these older plants appear dead from a distance, but close observation usually reveals a narrow band of green tissue, the plant's only link to life. Beyond the Pine Alpha grove, along a less-maintained road, is the Patriarch Grove, the greatest collection of bristlecone pines in the world.

A fourth expedition from Bishop took us to the southeast. Here, surrounded by the Inyo Mountains on the north and west, the colorfully banded Last Chance Mountains on the east, and the Saline Range on the south, is a large depression known as Eureka Valley.

At the south end of the valley is the dominating feature, a 700-foot-high mountain of sand nearly three miles long and up to one and one-half miles wide. These are the Eureka Dunes, home of many rare plants and animals, all living together and often dependent on each other in this fragile ecosystem.

The dunes were first recorded in July 1871 by one of the expeditions under the direction of Lt. G. M. Wheeler. Reports are that the plants collected during this expedition fell into a deep gully along with the pack mules and were never retrieved. The Wheeler Expedition reported that the "mules sank knee deep at every step."

The main dune covers an area of about four square miles, with an additional five square miles of lesser dunes. A sheet of sand surrounds all the dunes, except for a dry lake bed at the northwest corner. The dunes are composed of pale gold or cream-colored sand that is primarily quartz. Although the dunes are relatively stable, surface movements do occur that create ripple patterns and crests that build up and shift with the winds.

Mary DeDecker has spent several years studying the Eureka Dunes and surrounding areas during all seasons of the year. She reports great floral displays of desert mallow (*Sphaeralcea ambigua*), woolly desert marigold (*Baileya pleniradiata*), and the folded coldenia (*Coldenia plicata*), along with a beautiful white-flowered evening primrose (*Oenothera avita* ssp. *eurekensis*), a federally endangered plant.

Philip A. Munz and J. C. Roos discovered the Eureka Dunes evening primrose on September 18, 1954, growing in deep sand bordering the dunes. This plant does not occur on the dune slopes, but instead grows well away from them.

Oenothera avita ssp. *eurekensis* is a perennial which grows from a deeply penetrating root. The stems may stand up to two feet tall and are covered with white hairs. After the plants have shed their seeds, the stems often become buried by the shifting sands. Then, after the rainy season, if there is one, new growth of leaves may develop from the old fruiting tips that are buried under the sand. Large white flowers grow singly in the axils of the leaves. The flowers bloom in the evening, usually withering by midafternoon of the next day. They become somewhat reddish as they wither.

Until 1976, when the Bureau of Land Management closed the Eureka Dunes to off-road vehicles, the existence of the Eureka Dunes evening primrose was being jeopardized because of the trampling from them. As a result, the United States Fish and Wildlife Service declared *Oenothera avita* ssp. *eurekensis* to be endangered on April 26, 1978.

Eureka Dunegrass or *Swallenia alexandrae*.

Grass family. Perennial; stems rigid, to 1 foot. Leaves short, rigid, to 4 inches, with hard, pointed tips. Flowers inconspicuous, in narrow clusters. Season: April to June.

Did you know that some sand dunes sing? The Eureka Dunes do, one of the few in the world that exhibit this feature. The singing can sometimes be heard on the high sand ridges when small sheets of sand composed of rounded, polished grains, cascade across the slope. Mary Ann Henry, an amateur botanist who has studied the dunegrass that grows on the Eureka Dunes, describes the singing as "loud and rumbling, similar to the deep tones of a pipe organ with accompanying vibrations."

The dunegrass that Mary Ann Henry had been observing is the Eureka dunegrass, *Swallenia alexandrae*. This grass occurs in only four localities in Eureka Valley, where it is found in large clumps in deep sand. It grows on steep dune slopes to within 150 feet of the crest.

Annie Montague Alexander and Louise Kellogg were collecting companions in California and Nevada from 1938 to 1950. One of their trips was to the Eureka Dunes in May 1949 where Miss Alexander found the Eureka dunegrass. The dunegrass is a most unusual-looking species. It has stiff stems that stand a little more than one foot tall. Short rigid leaves with hard pointed ends, sharp enough to puncture the skin, occur all along the stem. The leaves are two to four inches long. Inconspicuous flowers in narrow clusters terminate the stems from April to June.

Eureka dunegrass can survive the dunes by forming extensive underground (or undersand) stems known as rhizomes. All along the rhizomes, new, erect aerial shoots can be formed. The dunegrass, once established in the dunes, stands in the way of wind-blown sand, causing the sand to pile up around the grass, forming rounded mounds. The mounding sand and the clumps of dunegrass are referred to as hummocks and, through time, may develop to as much as ten feet across.

One damage that off-road vehicles do to the dunegrass is to break the rhizomes apart with their churning action, tearing up all the essential water-absorbing fragile roots.

In addition to the Eureka Dunes evening primrose and the Eureka dunegrass, there are several other rare plants and animals known only from these dunes. Nearly a dozen kinds of flowering plants are restricted to this general area. More than 150 species of beetles have been found at the dunes, with at least 5 of them known from no other area. One kind of weevil, restricted to the Eureka Dunes, can survive only because of the shade provided by the Eureka dunegrass.

The plants and animals of Eureka Valley lived a reasonably undisturbed life in their remote corner of the world until the early 1960s, when motorcycles, dune buggies, and other off-road vehicles became popular. Huge "events" would be staged which damaged the living organisms and left the dunes as sandy trash piles. Through the continued efforts of conservationists in the area, the Bureau of Land Management closed the Eureka Dunes to off-road vehicles in October 1976. Still, "events" were held until the area had to be heavily patrolled. Now, off-road vehicle activity is at a minimum, fortunately. The Eureka dunegrass was listed as a federally endangered species on April 26, 1978.

Sonne's Barberry or *Berberis sonnei*.

Barberry family. Shrub; stems to 2 feet. Leaves alternate, compound, divided into 5 narrowly oblong, shiny, prickly leaflets to 3 inches. Flowers crowded in elongated clusters to 3 inches; petals 6, yellow. Season: April, May.

One of the most fascinating stories about a federally endangered species centers on the rediscovery of Sonne's barberry, *Berberis sonnei*. James B. Roof, for several years director of the Regional Parks Botanical Garden at Berkeley, California, was a leader in trying to relocate this barberry. His detailed account in the February 26, 1974, issue of *The Four Seasons* is summarized here.

Charles F. Sonne, an accountant for the Truckee (California) Lumber Company, collected a specimen of a small barberry on August 11, 1884, "on rocky banks of the Truckee River, Nevada County, California." The plant, originally misidentified as the rather common barberry, *Berberis aquifolium*, was placed in the Dudley Herbarium at Stanford University. It was there that Dr. Leroy Abrams of Stanford saw the specimen in 1934 and described it as a new species.

First interest in relocating the plant came in 1935, when the United States Forest Service initiated a California-wide project to collect seeds of native plant species for an erosion-control project along mountain roads. James Roof was the botanist on the project, whose job it was to germinate the seeds once they were collected.

Berberis sonnei was on the want list, but no one working on the project, including Dr. Roof, could relocate the species. Roof describes many trips between 1935 and 1972 along the Truckee River, where he and several agile teen-agers employed by him searched the rocky banks for the barberry, but to no avail. By then, Roof was ready to go on an all-out offensive to locate the barberry. He decided to organize an expedition. With the assistance of Dr. G. Ledyard Stebbins of the University of California, Mr. Gordon H. True, who had just completed a work on *The Ferns and Seed Plants of Nevada County, California,* and the California Native Plant Society, Roof planned an intensive search along the Truckee River for May 26, 1973. To coincide with the trip, an account of the upcoming foray, together with a photograph of the specimen collected by Sonne, was to appear in the local newspaper, the Sierra *Sun-Bonanza.*

On the front page of the May 23 *Sun-Bonanza,* a picture-framed story headlined "Botanical Detectives Hunt for Rare Shrub" appeared.

Roof arrived at Commercial Row in Truckee at nine o'clock on the morning of May 26 and found already assembled an array of people ready for the plant hunt. There were members of the Truckee-Donner Historical Society, the Truckee Garden Club, the advanced biology class of the Tahoe-Donner High School, the California Native Plant Society, the California Academy of Sciences, and various others from as far away as Monterey, Sacramento, San Francisco, and St. Helena.

Just before leaving for the hunt, a young man introduced himself to Dr. Roof as Nick Santamaria, the biology teacher at the high school. He in turn introduced a junior from his biology class, Miss Cathy Kramer. With Miss Kramer was a specimen of Sonne's barberry she had collected on May 1, 1973. More amazingly, she and Mr. Santamaria had correctly identified the specimen as *Berberis sonnei* three weeks before it had been publicized in the *Sun-Bonanza.*

Dr. Roof's story continues:

"Where did you get *that?*" I asked her.
"Down by the river. We can walk there from here."
Led by the high school set, the crowd steamed across the main street, past the east end of the railroad station, across the Southern Pacific Railroad's mainline tracks, through vacant lots and down to the Truckee River. There, on the rocky banks, Cathy Kramer had her picture taken as she sat on a man-made stone wall that had been invaded by stout plants of *Berberis sonnei.*

All told, two populations, occupying not more than ten feet of space, were located about five feet above the summer water level of the river. The plants were growing among and sometimes were overgrown by coarse vegetation.

Sonne's barberry is an evergreen shrub that grows no more than two feet tall. Each leaf is divided into five narrowly oblong, glossy green, prickly leaflets up to three inches long. The small yellow flowers are densely arranged in elongated clusters up to three inches long. Each flower has six petals. The flowers bloom in April and May.

This species was listed as federally endangered on November 6, 1979.

Raven's Manzanita or *Arctostaphylos hookeri* ssp. *ravenii*.

Heath family. Much-branched evergreen shrub lying on ground. Leaves alternate, simple, leathery, rounded to broadly elliptic. Flowers borne in clusters, very small. Season: February, March. (See color plate 37.)

California has the greatest collection in the world of wild manzanitas, a group of woody plants belonging to the genus *Arctostaphylos* in the heath family. Two of the most closely related manzanitas, as well as two of the rarest ever known, are the Laurel Hill manzanita and Raven's manzanita. Although the correct botanical names for these two have been kicked around, the California Native Plant Society calls the former plant *Arctostaphylos hookeri* ssp. *franciscana*, and the latter *Arctostaphylos hookeri* ssp. *ravenii*.

Dr. James B. Roof, former director of the Regional Parks Botanical Garden in Berkeley, California, has attempted to summarize the known historic locations for these two manzanitas. He says that both plants occurred together on serpentine rocks in only three places. These locations were the Laurel Hill Cemetery, the Masonic Cemetery, and Mount Davidson, all in San Francisco. Katherine Brandegee, who was curator of botany at the California Academy of Sciences in San Francisco from 1883 to 1894, was apparently the first person to collect these manzanitas, finding both kinds at the Laurel Hill Cemetery on April 2, 1907. Other collections of both plants were made by various botanists in 1918, 1923, and 1938. Neither one was found again for several years, and both were thought to be extinct.

Then, on April 10, 1952, Peter Raven, who later went on to the directorship of the Missouri Botanical Garden, discovered a manzanita sprawling on the ground in the Presidio, a military park in San Francisco near the entrance to the Golden Gate Bridge. Raven's plant was first called the Laurel Hill manzanita, but P. V. Wells later decided it was the other manzanita, which curiously enough had never been given a Latin name. Wells called the plant *Arctostaphylos hookeri* ssp. *ravenii*, or Raven's manzanita.

As it turns out, the specimen which Raven found at the Presidio, and which still occurs there today, is the only specimen known in the wild for Raven's manzanita. In addition, the Laurel Hill manzanita has not been found again anywhere and is presumed to be extinct.

The lone specimen of Raven's manzanita is an evergreen shrub that lies flat on the ground, forming a spreading mat several feet in diameter. The small leathery leaves are rounded or broadly elliptic. Small flowers less than one-fourth of an inch long bloom during February and March.

Although several cuttings made in the past from this plant at the Presidio are now growing as cultivated specimens in gardens, only a single wild plant is known today. Because of this, the United States Fish and Wildlife Service declared it an endangered plant on October 26, 1979.

Salt Marsh Bird's Beak or
Cordylanthus maritimus ssp. *maritimus.*
Snapdragon family. Annual; stems to 12 inches, gray-green. Leaves alternate, simple, very narrow, to 1 inch long, toothless. Flowers in crowded terminal clusters, creamy, to 1 inch long; petals 2-lipped, the upper lip with a yellow, beaklike tip. Season: May to July. (See color plate 38.)

Halophytes are plants able to grow in salt water. They have developed mechanisms which permit varying degrees of tolerance to salt. The majority of wetland plants in the United States are freshwater plants. If they were subjected to salt water, they would quickly perish. The reverse is also true. Saltwater-inhabiting plants cannot survive in freshwater conditions.

One of the rarest of these salt-tolerant plants is the salt marsh bird's beak (*Cordylanthus maritimus* ssp. *maritimus*). This plant is confined to a unique habitat known as the coastal salt marshes. The coastal salt marshes where this plant occurred at one time extended from northern Baja California, Mexico, to around Carpinteria in Santa Barbara County, California. Although these marshes will not take any prize for scenic beauty, they nonetheless provide a home for a fascinating assemblage of plants and animals.

A number of valleys have developed inland from the Pacific Ocean along the southern California coast. As the ocean current sweeps southward down the coast, an accumulation of sand develops at the mouth of these valleys. Often, enough sand is blown inland to form spits of sand that separate the lower ends of the valleys from the ocean. It is in these wet areas cut off from the ocean that the coastal salt marshes occur.

The plants and animals which live in these salt marshes have adapted themselves to surviving in an area where tides provide a regular washing with saltwater.

Although the salt marsh bird's beak once occurred in a number of the salt marshes from Baja California to Carpinteria, it is now restricted to a single large marsh south of San Diego and perhaps to two other marshes. Its decline is due mainly to the filling in of the coastal salt marshes by developers for one project or another.

To try to save the salt marsh bird's beak from extinction, the United States Fish and Wildlife Service listed it as federally endangered on September 28, 1978. Then, to add further protection to it and to three endangered birds, the United States Fish and Wildlife Service in 1980 acquired 505 acres of the last major undeveloped tidal estuary in southern California.

The salt marsh bird's beak, which is a member of the snapdragon family, is an eight- to twelve-inch-tall annual. Terminating each stem is a small, crowded cluster of creamy-colored flowers with a beaklike yellow tip. The flowers open from May to July.

Pointed Orcutt Grass or *Orcuttia mucronata*.

Grass family. Annual; stems spreading to erect, yellow-green, to 5 inches. Leaves alternate, narrow, long-tapering. Flowers inconspicuous, borne in crowded spikes to 2 inches. Season: May to July.

California has more different kinds of flowering plants than any other state in the United States. Although there are several factors which have contributed to this, one is certainly the great number of habitats in the state. One of the most interesting of these special habitats in California is the vernal pool, a habitat so unique that only in southern Oregon and in the Cape Province of South Africa can other vernal pools be found.

Most of California's vernal pools are in the Great Central Valley. Before settlement of this valley, these pools occurred in great

Eureka Dunegrass
Swallenia alexandrae

Pointed Orcutt Grass
Orcuttia mucronata

Sonne's Barberry
Berberis sonnei

McDonald's Rock Cress
Arabis macdonaldiana

numbers, but as the region was populated, the valley became a great agricultural area. In the process, countless vernal pools were destroyed.

Vernal pools exist in a wide range of sizes and depths. They occur in different soil types. They are of different ages. But they all have one thing in common. Beneath the shallow depression of the pool, a layer of compacted clay, known as a hardpan, has developed. This hardpan is impervious to water, so that during spring rains, the depressions of the pools fill with water that is unable to drain out below the hardpan. The result is the formation of a temporary body of water during early spring. As the water dries up from the summer's heat, the floor of the pool develops a network of deep cracks.

For a variety of reasons, the plants that live around the vernal pools form a special flora found nowhere else in the world. Many of the plants are annuals; some are extremely showy, such as the goldfields, popcorn-flowers, and downingias. Rarely are any aggressive, nonnative weeds found in or around the pools.

One group of annual grasses, belonging to the genus *Orcuttia*, is restricted to vernal pools. Orcuttias are strange in that they secrete a liquid over all of their structures. During the summer, the liquid hardens, becomes brownish and sticky, and emits a peculiar odor that can be detected some distance away. The strange odor still persists on specimens that were collected and dried many years ago.

There are several species of *Orcuttia* found here and there around the vernal pools in California, but one of them, the pointed Orcutt grass, *Orcuttia mucronata*, is exceptionally rare.

Dr. Beecher Crampton, an authority on *Orcuttia* in the Department of Agronomy and Range Science at the University of California at Davis, discovered this annual grass in a vernal pool in Solano County on August 1, 1958. The grass was found over a large portion of the floor of the pool. *Orcuttia mucronata* still lives in this vernal pool, and at no other place in the world. The pool is surrounded by agricultural land. A dirt road has been cut through the bed of the pool, but without any apparent effect on the grass. The precarious situation of this grass has been noted by the United States Fish and Wildlife Service, who listed it as a federally endangered species on September 28, 1978.

Each plant usually forms several yellow-green stems which sprawl out at first before becoming erect, reaching a height of about five inches. Long-tapering leaves, borne along the stems, sometimes surpass the flowering clusters. The groups of flowers are arranged in crowded, two-inch-long spikes. The flowers mature from May to July.

McDonald's Rock Cress or *Arabis macdonaldiana*.

Mustard family. Perennial; stems to 8 inches. Basal leaves spatula-shaped, to 1 inch; leaves on stem alternate, narrow, shorter. Flowers few on each stem, purple; sepals 4; petals 4, very narrow, to ½ inch long. Season: May, June.

The soil on the top of California's Red Mountain and on Little Red Mountain three miles south is unbelievably red. Walter and Irja Knight, who have surveyed the plants of these mountains, call it brick red. The soil is impregnated with rich amounts of nickel, and therein lies a problem with the plants that live on Red Mountain, because the Coastal Mining Company has long-range plans to remove as much of the nickel-containing soil as is economically feasible.

Red Mountain, which lies about fourteen miles in from the coast near Laytonville in Mendocino County, has an interesting assemblage of plants. Sixteen different kinds of flowering plants have their type locality on Red Mountain, that is, they were found for the very first time there. One of the new species from Red Mountain is a small, fragrant, purplish-flowered member of the mustard family which Alice Eastwood discovered on her first trip to the mountain on May 26, 1902. Miss Eastwood named the little plant she found *Arabis macdonaldiana*.

McDonald's rock cress was not seen again for forty years, although it must be admitted that apparently few botanists visited Red Mountain during that time. In 1942, Lincoln Constance and Reed C. Rollins went to the mountain and made the second collection of *Arabis macdonaldiana*. Then, during mid-July 1969 the Knights and friends Roman Gankin and Richard Hildreth surveyed Red Mountain and located McDonald's rock cress again.

Although *Arabis macdonaldiana* grows on Red Mountain at elevations between 3,500 and 4,000 feet, it apparently does not occur in the red soil of nearby Little Red Mountain. In fact, it was known only from Red Mountain until just recently, when a small population was found in Oregon. Several colonies of a mustard which may be McDonald's rock cress have been found along tributaries to the Smith River in Del Norte County, California. A study to determine the identity of this mustard is under way by biologists from Humboldt State University.

This mustard is a dwarf perennial whose unbranched stems reach only a height of eight inches. There is a ring of spatula-shaped leaves at the base of the plant, while a few much smaller and narrower leaves grow along the upper part of the stem. None of the leaves is more than one inch long.

A few purplish flowers terminate the stem in May and June. They are composed of four very small sepals and four narrow petals less than one-half inch long and which taper to the base.

The extremely limited distribution of this species and the threat from nickel mining enabled the United States Fish and Wildlife Service to list McDonald's rock cress as endangered on September 28, 1978.

Santa Barbara Island Live-Forever or *Dudleya traskiae*.

Sedum family. Succulent perennial. Basal leaves crowded in rosettes, thick, to 6 inches. Flowers several on 1-foot stalk, bright yellow; petals to ½ inch. Season: April, May.

Off the coast of southern California, from Santa Barbara to San Diego, lie a series of islands which have attracted the attention of biologists because of the many species of plants and animals confined to those islands.

There are eight of these islands, which are generally referred to as the Channel Islands. The northern group of these, called the Northern Channel Islands, include San Miguel, Santa Rosa, Santa Cruz, and Anacampa. The Southern Channel Islands are made up of San Clemente, Santa Catalina, Santa Barbara, and San Nicholas. These islands range in size from Santa Barbara, which is barely one square mile, to Santa Cruz, which is nearly one hundred square miles. Anacampa is the closest of the Channel Islands to the mainland, only thirteen miles away, while San Nicholas is sixty-one miles from mainland California.

In summarizing the unusual nature of the flora, Dr. Peter Raven in 1967 noted that there is a total of seventy-six different kinds of flowering plants confined to one or more of these islands and which do not occur on the mainland. Eighteen of these are restricted to just a single island.

Santa Barbara Island, the smallest of the Channel Islands, is now a part of the Channel Islands National Monument, administered by the National Park Service. The island, somewhat triangular in shape, is located thirty-eight miles off the California mainland. The island is mostly surrounded by steep, rugged sea cliffs, some rising as much as 500 feet above the ocean. Numerous caves and coves add dramatic beauty to the island.

The native vegetation of Santa Barbara Island was at one time remarkable. The island was famous for its giant coreopsis, or tree sunflowers. But after the island was subjected to farming practices, including grazing and intentional burning, much of the native vegetation became replaced by weedy, introduced plants. In 1967, Dr. Raven reported that there were only forty native species on Santa Barbara Island. One of these, however, was a very rare succulent species called *Dudleya traskiae*, the Santa Barbara Island live-forever.

This species was first collected in 1901 by a woman from the islands, Blanche Trask. Mrs. Trask, a resident of Santa Catalina Island, collected plants on several of the Channel Islands around the turn of the century. Her expeditions to Santa Barbara Island in May 1901 and 1902 were the first thorough studies on the island by a botanist.

Mrs. Trask had found the live-forever on exposed cliffs and rocky slopes. At about the same time that farmable areas of Santa Barbara Island were being cleared, New Zealand red rabbits were introduced to the island. When Dr. Reid Moran of the San Diego Museum visited the island in 1941 with M. B. Dunkle, only two populations of *Dudleya traskiae* could be found, evidently having been grazed by the introduced rabbits. By 1949 only one population could be located, and in 1970 even it could not be found. Ralph Philbrick of the Santa Barbara Botanical Garden noted in his 1972 article "The Plants of Santa Barbara Island, California" that this species was probably extinct in the wild, although it was being cultivated in several gardens.

In May 1975 five living plants of *Dudleya traskiae* were found by Philbrick on the east side of Santa Barbara Island. Since these plants were subjected to nibbling and grazing by the introduced rabbit population, National Park Service biologists went on a rabbit hunt to rid the island of these succulent-eating pests. Today, the Santa Barbara Island live-forever is known from three small populations in three canyons and one large colony on an inaccessible, vertical cliff.

Dudleya traskiae is a perennial with rosettes of twenty-five to thirty-five crowded, succulent leaves from two to six inches long. When the plant is about to flower, it sends up a stalk nearly one foot tall which bears many bright yellow flowers.

The vulnerable nature of this species enabled the federal government to list it as endangered on April 26, 1978.

San Clemente Island Broom or *Lotus dendroideus* ssp. *traskiae*.
Pea family. Perennial; stems to 3 feet. Leaves alternate, compound, divided into 3 oblong leaflets to ⅓ inch. Flowers usually 5 in an umbrellalike cluster, sweet-pea-shaped, to ½ inch, yellow tipped with red. Season: February to August.

Blanche Trask spent much of her life exploring the Channel Islands, where she found several plants new to science. Some of these have been named in her honor. One of the islands which she loved to visit was San Clemente, which she once described as an "amethystine beauty, like an Indian arrow-head, tipped with shining stretches of sand, enshrined by the white arms of the sea."

San Clemente is the southernmost of the Channel Islands. It lies fifty miles south-southwest of San Pedro and about sixty-four miles west-northwest of San Diego. San Clemente Island extends for nearly twenty-one miles, with its widest part about four miles across. The total land mass is nearly fifty-six square miles. Near the center of the island is the highest elevation, at 1,965 feet.

Mrs. Trask continues with her description of the island: "a rolling upland strewn with jagged volcanic rocks . . . reaches its greatest altitude on the north [east] coast—a coast gashed by precipitous and bold gorges. The south [west] coast rises from the sea with perpendicular walls, fifty to three hundred feet high. . . ." It is in these canyons and on the rocky cliffs that most of the unusual plants occur.

Dr. Peter Raven, now director of the Missouri Botanical Garden, published *A Flora of San Clemente Island, California*, in 1963. In his work, Raven noted that there are 233 different kinds of native plants that grow on San Clemente Island. Of these, 14 are known only from this island, and 4 of these became the first plants ever to be listed as federally endangered.

Blanche Trask searched for plants on San Clemente Island at least three times. Her first visit was in October 1896, and her second was during October 1902. Then, in 1903, she lived on the island for three months during the spring of the year. It was on her first trip that she found a bushy, yellow-flowered member of the pea family which later was to be named *Lotus dendroideus* ssp. *traskiae*, the San Clemente Island broom.

The San Clemente Island broom grows primarily on grassy slopes or on rocky cliffs. Although grazing by feral animals on San Clemente Island for many years has greatly reduced much of the native plant life, these wild animals apparently graze very little on the San Clemente Island broom. The main threat to this plant seems to be road building and construction by military personnel, since the entire island is owned by the United States Navy. Several pipelines cross the habitat of *Lotus dendroideus* ssp. *traskiae*, and roads to gain access to power lines have contributed to a decline of this plant.

Another problem is that this plant may be in danger of losing its identity, since it is known to hybridize with another kind of *Lotus* found on the island.

The United States Fish and Wildlife Service listed this plant as endangered on August 11, 1977.

The San Clemente Island broom is a bushy perennial that grows to a height of about three feet. Most of its leaves are divided into three oblong leaflets which are one-fourth to one-third inch long. There are usually about five yellow flowers borne in an umbrellalike cluster. The petals are yellow tinged with red and about one-half inch long.

San Clemente Island Bushmallow or
Malacothamnus clementinus.

Mallow family. Shrub to 3 feet. Leaves alternate, simple, lobed, star-shaped, hairy. Flowers crowded into spikes, rose pink; petals 5, to ¾ inch. Season: March to August.

When goats were first introduced to San Clemente Island in the early 1800s, it meant the beginning of the end for several species of plants and a strong reduction in most other native plants on

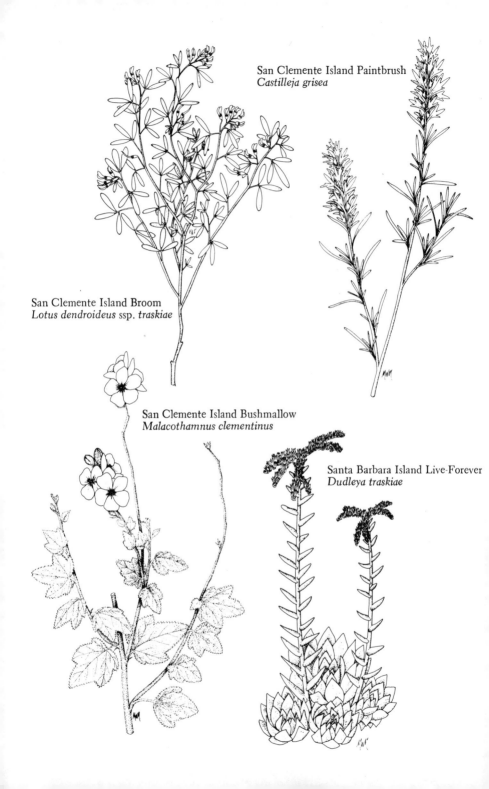

San Clemente Island Paintbrush
Castilleja grisea

San Clemente Island Broom
Lotus dendroideus ssp. *traskiae*

San Clemente Island Bushmallow
Malacothamnus clementinus

Santa Barbara Island Live-Forever
Dudleya traskiae

the island. In 1877 the Department of Commerce leased the island to the San Clemente Sheep and Wool Company. Now there were both goats and sheep to intimidate the plant life of the island.

After a period of fifty-seven years and countless numbers of plants eaten, digested, and expelled, the sheep were removed from San Clemente Island in 1934. But the goats remained. What a goat can do to a plant in a short time shouldn't happen to a dog-bane, or a dogwood, or a bushmallow, for that matter.

The bushmallow that has provided food for the goats is *Malacothamnus clementinus*, the San Clemente Island bushmallow.

Philip Munz, author of *A California Flora*, who served for many years as the director of the Rancho Santa Ana Botanic Garden, spent five days on San Clemente Island in 1923 as one of a party of nine. According to Peter Raven in his *Flora of San Clemente Island, California*, "the party landed at Wilson Cove, circled the north end dunes, and the next day set out by wagon for the south, camping near Lemon Tank and then working the canyons on both sides of the island to the vicinity of Middle Ranch and Thirst." It was on this trip that Munz discovered the San Clemente Island bushmallow "on walls of canyon running into the sea."

In 1975, about eight more plants of the bushmallow were found in a second canyon on the island. When Howard A. Ferguson, a botanist employed by the United States Navy at San Clemente Island, sought to relocate Munz's original population, he was able to find eight to ten clumps of fifteen to twenty individuals.

At both known localities on San Clemente Island this species grows in relatively inaccessible locations that even the goats would have difficulty reaching.

In order to give maximum protection to the San Clemente Island bushmallow, the United States Fish and Wildlife Service declared *Malacothamnus clementinus* to be endangered on August 11, 1977.

As part of the management plan developed by the United States Navy, which owns the island, the California Department of Fish and Game, and the United States Fish and Wildlife Service, efforts have begun to reestablish the natural ecological conditions of the island. To do this, it became necessary to remove the goats from the island. By trapping and herding, the population of goats was reduced from 12,500 in 1977 to 1,500 in 1979. Those that remained took refuge in inaccessible canyons where trapping was ineffective. In order to eliminate the remaining goats, an aerial shooting program, to last for one week in June 1979, was approved. However, just a few days before the aerial shoot was to take place, the Fund for Animals filed suit to prevent it. During the delay brought on by the suit, the spared goats continued to reproduce rapidly. Then, on January 23, 1980, the United States District Court for the Central District of California ruled that 60 percent of all feral animals on the island were to be counted and live-trapped within a ninety-day period and removed from the island by barge. The remainder would have to be taken from the island within one year.

The San Clemente Island bushmallow is a rounded shrub with numerous branches up to three feet long. Lovely hibiscuslike rose-pink flowers are crowded together in spikes that arise from the axils of the upper leaves. The flowers bloom from March to August.

San Clemente Island Paintbrush or *Castilleja grisea.*
Snapdragon family. Perennial; stems slightly woody, to 2 feet, gray-hairy. Leaves alternate, simple, narrow, without teeth. Flowers subtended by lobed bracts; petals united, yellow, to 1 inch. Season: May to August.

Among the most gorgeous wildflowers of the western United States are the paintbrushes of the genus *Castilleja*. These members of the snapdragon family are widespread in the western United States, where their flowers range in color from yellow to orange to red and all hues in between. But there is only one paintbrush on San Clemente Island, and it is in danger of becoming extinct.

The San Clemente Island paintbrush, *Castilleja grisea*, was discovered by T. S. Brandegee when he went to San Clemente Island in 1894 as part of a biological expedition of the International Boundary Commission. Brandegee, however, did not realize he had found a new species, nor did Blanche Trask when she found the same plant on her visit to the island in 1896. In reporting the results of her foray, Mrs. Trask noted that "a strange *Castilleia* [sic] here flourishes, with rich canary-colored bracts shrubby, two to four feet tall."

This new, unnamed paintbrush was collected twice more, once by Mrs. Nell S. Murbarger in 1936 and once by Francis H. Elmore with a group from the Allen Hancock Foundation in February

1939. Then, on April 2, 1939, Meryl B. Dunkle found it and named it *Castilleja grisea*. Dunkle was with the Los Angeles Museum, which was surveying the biology of the Channel Islands. When Elmore made his collection in 1939 from the lower slope bluffs on the southeast coast of the island, he considered the plant to be common. In 1963, Peter Raven noted that *Castilleje grisea* was occasional on the southeast coast and very rare in canyonsides elsewhere. But by 1978, Thomas Oberbauer, a frequent visitor to the island, stated that on numerous trips there he had seen only three plants, all on the canyon walls. They were so inaccessible that it took Oberbauer more than an hour to get to and back from one specimen that was only seventy feet above him on the canyon wall. The United States Fish and Wildlife Service listed this species as endangered on August 11, 1977.

R. Mitchel Beauchamp and Howard L. Ferguson of the Pacific Southwest Biological Services estimated late in 1979 that the total population of the San Clemente Island paintbrush was about 450 plants, but that they were threatened by goat predation.

Castilleja grisea is a somewhat shrubby plant growing about two feet tall. The flowers, which are dull yellow, bloom during late spring and summer.

San Clemente Island Larkspur or *Delphinium kinkiense*.
Buttercup family. Perennial; stems to 1½ feet. Leaves alternate, simple, deeply lobed. Flowers 8–10 in a cluster, white and pale violet, nearly 1 inch long. Season: March, April.

Although a number of people have explored San Clemente Island for plants, beginning in 1885 with William Scrugham Lyon and Rev. Joseph C. Nevin, two amateur botanists from Los Angeles, it was not until 1939 that the San Clemente Island larkspur was discovered, and not until 1969 that it was named a new species.

Meryl B. Dunkle first found this plant in a high grassy area on April 6, 1939, but his collection was identified as another kind of larkspur. Then, on March 18, 1967, R. Mitchel Beauchamp found the larkspur again on grassy slopes at the head of a canyon. When Philip Munz examined Beauchamp's specimen, he decided it was a new species and called it *Delphinium kinkiense*. Kinkiense, incidentally, is derived from "kinki," the Gabrielmo Indian name for San Clemente Island.

It is ironic that this species, which is grazed upon by goats, should have been found by Beauchamp at an area on the island called Nanny. This species is limited to seven grassland sites on San Clemente Island, with a total population of fewer than twenty individuals. The threats to the existing plants include grazing by goats, rooting by feral pigs, possible fires, and simply crushing and trampling, since the species occurs on rather level terrain. The United States Fish and Wildlife Service listed *Delphinium kinkiense* as endangered on August 11, 1977.

The San Clemente Island larkspur is a perennial with erect stems up to one and one-half feet tall. Several deeply lobed leaves are borne along the stems. Each cluster of flowers has eight to ten blossoms which are a combination of white and pale violet. The flowers, which bloom in March and April, are slightly less than one inch long.

San Diego Mesa Mint or *Pogogyne abramsii*.
Mint family. Annual; stems to 6 inches. Leaves opposite, simple, very narrow, to 1 inch, usually absent at flowering time. Flowers few on each stem, to 1 inch; petals 2-lipped, reddish-purple mottled with white. Season: April to June.

There is an area north and east of San Diego where vernal pools are found on mesas or terraces. Most of the mesas are flat, except where they have been cut by gulleys. During the rainy season, many pools form in depressions on the mesas, supporting a unique flora distinct from that of the surrounding areas.

Probably the first person to study the vernal pools north of San Diego was Charles Russell Orcutt, who visited them in 1895 and 1897. When LeRoy Abrams of Stanford University went to the vernal pools north of San Diego in the early 1930s, he discovered a small mint with reddish-purple flowers. Abrams, who is best known for his four-volume *Illustrated Flora of the Pacific States*, did not realize that he had found a new species. But a few years later, when John Thomas Howell of the California Academy of Sciences studied the mint that Abrams had collected, he recognized it as a new species, calling it *Pogogyne abramsii*, the San Diego mesa mint.

Edith A. Purer, a high school teacher in San Diego, made an

extensive study of the vernal pools of San Diego County in 1937. In that study, Purer noted that *Pogogyne abramsii* was "commonly present in the vegetation of the vernal pools."

If *Pogogyne abramsii* was common in 1937, it is very uncommon today, due to human disturbance of the habitat. Several populations have been lost because of housing developments, off-road vehicles, highway widening, and illegal dumping. Other colonies were wiped out when the mesas on which they grew were converted to agriculture. The San Diego mesa mint now occurs on only three mesas. In an effort to salvage what is left of this species, the federal government listed *Pogogyne abramsii* as endangered on September 28, 1978.

The San Diego mesa mint is a tiny annual that grows to a height of six inches. Several pairs of opposite leaves, each narrow and no more than one inch long, are borne along the stem. Only a few flowers are formed on each stem. Each flower, which is two-lipped and about one inch long, is reddish-purple with some white mottling. Often, when this species flowers in April, May, and June, most or all of the leaves will have dropped off. The San Diego mesa mint has a strong, minty odor when crushed.

Contra Costa Wallflower or
Erysimum capitatum var. *angustatum.*
Mustard family. Coarse biennial; stem to 2½ feet. Leaves alternate, simple, narrow, to 6 inches. Flowers showy, yellow; sepals 4; petals 4, to 1 inch. Season: March to July.

If anyone who saw the Antioch Dunes a century ago could return to the area today, he would scarcely recognize the region. What used to be a two-mile stretch of river dunes, some as high as 115 feet, has been reduced to mounds no more than 30 feet tall. The dunes lie on the south side of the San Joaquin River east of the town of Antioch, California.

A continuous series of activities, beginning at least as early as 1855, has degraded and nearly done away with the dunes. Alice Q. Howard of the University of California at Berkeley, who heads a group trying to save the Antioch Dunes and their plants and animals, has reported that a pottery works opened nearby in 1865 and two brickyards were established before the turn of the century. Active sand mining began in 1921, and the dunes were lowered bit

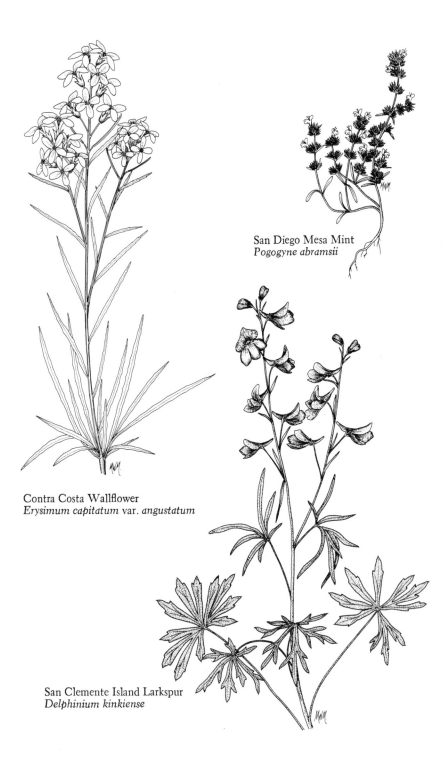

San Diego Mesa Mint
Pogogyne abramsii

Contra Costa Wallflower
Erysimum capitatum var. *angustatum*

San Clemente Island Larkspur
Delphinium kinkiense

by bit for use in asphalt and concrete mixes. What rare plant and animal life that wasn't actually mined away was crushed and severely disturbed by massive pieces of equipment. And believe me, many of the kinds of plants and animals that call the Antioch Dunes their home are rare ones.

J. Burtt-Davy, who botanized the dunes in 1895, wrote that "these sand-hills are brilliant with flowers and, being left uncultivated, form fine botanizing ground." The plants that weren't destroyed by mining operations eventually lost out to aggressive weeds which invaded the sand after heavy disturbance.

One of the rarest flowering plants confined to the Antioch Dunes is a showy member of the mustard family called the Contra Costa wallflower, *Erysimum capitatum* var. *angustatum*.

This plant was first collected by Edward L. Greene of the University of California. Greene was plant-hunting on the dunes on April 14, 1887, when he came upon the Contra Costa wallflower along the sandy banks of the San Joaquin River.

Erysimum capitatum var. *angustatum* is a coarse biennial that grows to a height of two and one-half feet. Several very narrow leaves are produced, most of them clustered near the base of the stems. The largest of these may reach as long as six inches. Several gorgeous flowers are borne at the ends of some of the stems from March to July. Each flower has four yellow petals that are nearly one inch long.

On April 26, 1978, the United States Fish and Wildlife Service listed the Contra Costa wallflower as federally endangered because many of the populations had been destroyed by mining operations. In justifying the listing, it was noted that fewer than 200 plants actually survive. But since the Antioch Dunes were privately owned, federal laws regulating plant protection did not apply. More had to be done. After negotiating with two parties who were the primary owners of the dunes where most of the Contra Costa wallflowers occur, the United States Fish and Wildlife Service announced on March 6, 1980, that the federal government had purchased two tracts of sand, totaling slightly more than fifty-five acres. This landmark action was the first time that private property was purchased by the federal government to protect endangered plant life.

But now another problem arose. Since the fifty-five acres were now owned by the federal government, some persons, particularly those with off-road vehicles, felt that the protected dunes had become public property. The San Francisco Bay National Wildlife Refuge, whose personnel administer the dunes, has had to take steps to protect the area from vandalism and further destruction.

Antioch Dunes Evening Primrose or
Oenothera deltoides ssp. *howellii.*
Evening primrose family. Biennial; stems to 2 feet. Leaves alternate, simple, lance-shaped, toothed or lobed, grayish. Flowers showy, to 3 inches across; petals 4, white; stamens 8. Season: March to May.

In addition to the Contra Costa wallflower, some other flowering plants and several kinds of insects which live on the Antioch Dunes are in grave danger of becoming extinct.

The insects of the dunes were first studied in 1931 by scientists from the University of California at Berkeley and Davis and from the California Academy of Sciences. Among these early entomologists was the well-known John Adams Comstock. Comstock made several discoveries, one of which was a butterfly known as Lange's metalmark butterfly. This insect, found only on the Antioch Dunes, can survive only if a particular kind of wild buckwheat is present for the larvae to feed upon. Since the buckwheat has been severely reduced in numbers (although not low enough to be listed as endangered at this time), Lange's metalmark butterfly has come close to extinction. On June 1, 1976, this butterfly was listed as an endangered species.

Besides Lange's metalmark butterfly, twenty-three other kinds of insects were initially discovered at the Antioch Dunes. Ten of these have never been found anywhere else; two apparently are already extinct.

One of the endangered plants that lives on the Antioch Dunes is a showy-flowered evening primrose. Although this plant had been collected as early as 1879, it was not until 1949 that it was given a Latin name. Today, the plant is called *Oenothera deltoides* ssp. *howellii.* At the dunes,, this evening primrose occurs in loose sand and stabilized dunes along the margin of the San Joaquin River.

Although sand mining has wiped out much of the habitat for the Antioch Dunes evening primrose, the periodic rototilling of the area for fire-control purposes has permitted the introduction of weedy species which eventually crowd out the evening primrose. The federal government listed it as endangered on April 26, 1978.

Since one of the necessary conditions for the survival of the Antioch Dunes evening primrose is the availability of freshly deposited sand, Alice Howard and her committee to save the Antioch Dunes are working to have the Corps of Engineers deposit sand onto the area.

Oenothera deltoides var. *howellii* is a biennial with somewhat drooping branches up to two feet tall. Several large flowers up to nearly three inches broad open from March to May. Each flower has four petals and eight stamens.

MacFarlane's Four-O'clock or *Mirabilis macfarlanei.*
Four-o'clock family. Perennial; stems to 3 feet. Leaves opposite, simple, nearly round, toothless, to 3 inches. Flowers 4–7 in a cluster, subtended by greenish-purple bracts; petals rose-purple to pink, funnel-shaped, to 1 inch. Season: May.

The Snake River, flowing northward as it forms the boundary between Idaho and Oregon, has carved the deepest gorge in North America. North from Copperfield, past the Hell's Canyon Dam and Pittsburgh Landing to the junction with the Salmon River, the Snake River is dwarfed by the steep walls of basaltic rock on either side. Some of the area is virtually inaccessible, while other parts can be traversed only by rugged hikers.

During the 1930s, E. B. MacFarlane piloted boats up and down Snake River. He had noted during his excursions an attractive pinkish-flowering herb growing at one place along the river on the Oregon side. Thus, when a botanical expedition led by Lincoln Constance and Reed C. Rollins came to the Snake River Canyon on May 15, 1936, MacFarlane had someone to show his unknown plant to. Constance and Rollins found that the boatman's plant was a new species of four-o'clock, upon which they bestowed the name *Mirabilis macfarlanei*, MacFarlane's four-o'clock.

Since the area is nearly inaccessible and the plant is very rare, little has been seen of this plant since. In 1947, a small group of MacFarlane's four-o'clocks was discovered along the Salmon River in Idaho, about ten miles due east of the original Snake River location. Today, this species is found in only a few locations in Oregon and Idaho, along the Snake and Salmon rivers.

Because of the very conspicuous nature of the plant, and because the areas where it occurs are receiving more and more visitor use, the United States Fish and Wildlife Service declared MacFarlane's four-o'clock as an endangered species on October 26, 1979.

The habitat for *Mirabilis macfarlanei* is in the open, either on steep sunny slopes or on gravel bars with full exposure to the sun.

MacFarlane's four-o'clock is a deep-rooted perennial which grows in clumps. The stems, which may grow to a height of three feet, bear several pairs of opposite, nearly round, toothless leaves about two to three inches long. From the upper parts of the plant are borne several clusters of bright rose-purple to pink flowers.

Osgood Mountain Milk Vetch or *Astragalus yoder-williamsii.*
Pea family. Tufted perennial; stems very short, some only with sharp-pointed leaf stalks. Leaves alternate, compound, divided into 7–19 tiny, folded leaflets. Flowers 2–8 per stem, sweet-pea-shaped, white with faint pink stripes, to ½ inch long. Pods elliptic, to ⅓ inch. Season: July, August.

One of the 1979 amendments to the Endangered Species Act gave authority to the United States Fish and Wildlife Service to issue emergency endangered status to any plant that it felt might become extinct before the normal process of listing a species as federally endangered could become enacted. The emergency listing, which would provide protection to the plant from whatever outside forces might be threatening the species, could be in effect for not more than 240 days. After the 240 days had elapsed, the United States Fish and Wildlife Service either would have to make efforts to list the species as endangered through the normal process, or would have to drop the listing.

The first use of this emergency listing process came on August 13, 1980, when a rare and little-known legume from southwestern Idaho and northwestern Nevada was declared endangered on an emergency basis.

The species which received the first emergency listing was the Osgood Mountain milk vetch, *Astragalus yoder-williamsii*, a plant that had been named only about a year before. Michael Yoder-

Williams, who manages the Sagehen Creek Field Station in the eastern Sierra Nevadas, had found this low-growing perennial on July 11, 1979, in the Osgood Mountains in Humboldt County, Nevada. The plants were growing on exposed ridge crests and flat plateaus of decomposed granite or sandy soil at an elevation of 7,120 feet above sea level. The original specimen was sent to Dr. Rupert C. Barneby, the world's authority on milk vetches at the New York Botanical Garden, who confirmed that Yoder-Williams's plant was a new species and accordingly gave it its Latin name. As it turned out, a very small population of this same species had been found in 1977 across the state line in Owyhee County, Idaho, by Sarah Richards.

In spring 1980, the United States Fish and Wildlife Service estimated that there were about 500 plants of this species living in Nevada and 10 in Idaho. At both sites where *Astragalus yoder-williamsii* was growing, mining claims for tungsten and gold ore had recently been made, and the Nevada station seemed in imminent danger. The emergency listing stopped any prospecting for the time at the Nevada site, but mining operations continue within a mile of the endangered plant. There is even a road that runs right through the major population of this species. On April 15, 1981, the emergency listing for *Astragalus yoder-williamsii* expired, and the species is now going through the normal process for federal listing.

This species is a dwarf perennial that grows in tufts. At the tip of some of the stems are two to eight flowers that do not overlap each other. The white petals have faint pink stripes. The longest of the petals is about one-fourth inch. The elliptic pods are about one-third inch long, but only one-tenth inch wide. The flowers bloom during July and early August.

Cobra Plant or *Darlingtonia californica.*

Pitcher plant family. Perennial; leaves modified into tubular pitchers with hood and forked, tonguelike projection. Flower solitary on a leafless stalk; sepals 5, green; petals 5, crimson, remaining closed. Season: April to June. (See color plate 39.)

My first trip to Oregon came shortly after I had embarked upon my professional career as a botanist. I was surprised one day as I was driving north on Highway 101 near Florence, Oregon, to come upon a roadside sign announcing a botanical wayside ahead. Having never seen a highway sign like that before (nor since, I might add), I had no idea what to expect.

I pulled the station wagon into the parking lot and climbed onto a wooden boardwalk that led out across a boggy area. Suddenly, before my eyes, was the most fantastic sight I had seen— hundreds of plants standing about two feet tall, each with an arched hood and a forked "tongue," looking for all the world like a serpent poised for the strike! I was seeing my first cobra plants, insectivorous organisms that are found in a limited region along the Oregon Coast and across into some southern Oregon and northern California mountains.

The cobra plant, *Darlingtonia californica*, is related to the pitcher plants of the eastern United States. Some of the leaves are modified into elongated, tubular pitchers which broaden out into a hood at the top. Beneath the hood is an opening, or trap, through which insects may pass. Projecting downward from the hood is a broad, forked, colorful structure referred to as the "fish-tail." The fishtail secretes nectar which attracts insects that crawl up the fishtail and into the opening of the hood, which is also lined with nectar.

The rounded part of the hood is provided with small, transparent "windows" which permit light to shine through. The insect, once in the trap, is attracted to the light, thus becoming further removed from the opening into the hood, its only escape route. Unable to leave the hood by way of the windows, the insect finally falls downward into the tube, dropping into water that has accumulated. Downward-pointing hairs on the sides of the tube prevent the insect from crawling back out.

Although the liquid in the pitcher of the cobra plant does not contain digestive enzymes as it does in pitcher plants, it does have bacteria in it which act upon the prey, ultimately causing it to decompose.

Cobra plants possess extensive underground stems which send up shoots at rather close intervals. This accounts for the dense colonies of plants that sometimes are found.

Despite this seemingly effective means of vegetative reproduction, *Darlingtonia californica* also produces flowers and fruits. A single flower is borne at the end of a long, leafless stalk that has a strong crook in it just before the attachment of the flower. Each flower has five narrow, green, spreading sepals and five crimson petals which come together at a point, never fully opening. The closed petals conceal a large bell-shaped ovary surrounded by twelve to fifteen stamens.

The original discovery of the cobra plant was made by William D. Brackenridge in 1841 in California's upper Sacramento Valley. Brackenridge was an assistant botanist for Capt. Charles Wilkes's United States Navy Exploring Expedition en route from the Willamette Valley to the Sacramento River. I would love to have seen the look on Brackenridge's face when he saw his first cobra plant.

There was considerable interest in this remarkable plant following its discovery. Norden Cheatham of the California Native Plant Society, writing recently in that organization's journal *Fremontia*, noted that in 1853, Asa Gray, the famous eastern botanist, wrote, "let our Californian readers take notice, that a small box of [cobra plant] roots, delivered alive in Boston, New York, or London, would be pecuniarily as valuable as a considerable lump of gold, and would furnish a handsome and highly curious acquisition to our gardens."

Having seen *Darlingtonia californica* near the Oregon coast, I was surprised to learn that the plant also grows in the Klamath and Sierra Nevada mountains of Oregon and California. Despite the great differences in elevation, most cobra plant colonies seem to have two basic growth requirements. They live in serpentine soils (how appropriate for a plant that looks like a serpent) in bogs that have a continuous or near-continuous supply of running water from upslope seepage.

Although *Darlingtonia californica* is not threatened with extinction at the present time, it is always the subject of exploitation because of its unique form.

Sand Food or *Ammobroma sonorae.*

Lennoa family. Nongreen, mushroomlike succulent; underground stems to 5 feet, above-ground stem with a flattened "cap" on the surface of sand. Flowers arranged in a ring over the "cap," small, purple; petals 5, united. Season: April, May.

"Colonel Gray at the campfire with some sand food." Sound like a statement from the parlor game "Clue"? Actually it describes

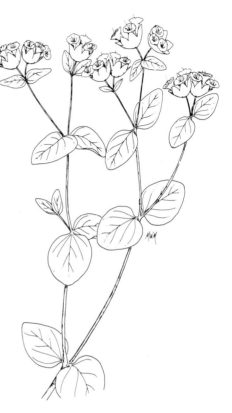

Antioch Dunes Evening Primrose
Oenothera deltoides ssp. *howellii*

MacFarlane's Four-O'clock
Mirabilis macfarlanei

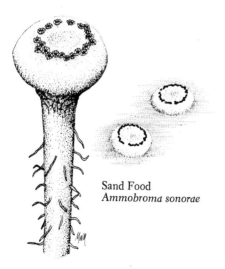

Sand Food
Ammobroma sonorae

Osgood Mountain Milk Vetch
Astragalus yoder-williamsii

a little scene that took place in 1854 in the great sandy area south and west of Yuma, Arizona.

Col. Andrew B. Gray had been hired by the federal government to survey potential railroad routes to the Pacific. On that day in 1854 he was traveling through the sand region inhabited by the Papago Indians of Arizona and Sonora, Mexico. A young brave had served at Gray's guide and had invited Gray to eat with the Papago chief that evening around the campfire. Gray noted that the chief roasted the stems of a plant called sand food and that it was "luscious, resembling the sweet potato in taste, only more delicate."

The sand food is an unusual nongreen flowering plant that looks like a mushroom but bears a ring of flowers on the "cap." It has been given the name *Ammobroma sonorae*, and it occurs in the pure sand that covers many square miles on either side of the United States–Mexican border between Yuma, Arizona, and Imperial, California.

The body of the plant is composed of an underground pedestal-like stem surmounted by a broad, flat, buttonlike cap that lies flat on the surface of the hot, dry sand. Small purple flowers with five united petals form a ring over the surface of the button.

Since the sand food plant is nongreen and lacks chlorophyll, it cannot manufacture its own food. In order to obtain nutrients, it must parasitize some host plant that grows in the sand. To do this, the plant elongates its stem beneath the sand until it comes into contact with the roots of a suitable host. The sand food plant will not attach itself to the roots of just any plants, but only to certain species. Wayne P. Armstrong of Palomar College in San Marcos, California, has excavated the underground system of *Ammobroma sonorae* and found one stem that had reached a length of five feet before it connected to the host.

One another occasion, Franklin A. Thackery discovered that 106 of the buttons he found in an area were all attached to the same extensive underground stem, and the whole plant weighed forty-five pounds!

Sand food is a remarkable plant, and it grows in a remarkable habitat. The greatest concentration of it is in the Algodones Dunes, a California Natural Area in the vicinity of Glamis. The Algodones Dunes extend in a continuous fifty-mile-long strip, with some of the dunes rising as much as 300 feet above the surrounding terrain.

This plant occurs sporadically, seemingly in response to moisture availability. Following winters with abundant rainfall, the sand food seems to be more abundant.

Gary Nabhan of the University of Arizona's Office of Arid Land Studies, who has worked extensively with *Ammobroma sonorae*, has found that in May, when the plant was in flower and the air temperature was between 98° and 100° F., the temperature at the surface of the sand next to the flowers was 136° to 141° F., while the temperature of the sand one foot deep was 77° to 80° F.

Ammobroma sonorae has been found in southeastern California, southwestern Arizona, and adjacent Sonora, Mexico. More studies are needed to determine just how common or how rare this species is. The United States Fish and Wildlife Service is reviewing its status.

Vanishing Species
of Hawaii and Alaska

Although Hawaii and Alaska have little in common, other than their distance from the contiguous forty-eight states, they are included in this chapter. Hawaii is noted for its diverse flora, which is composed of many species already extinct and a great number close to extinction.

The chain of volcanic islands that make up Hawaii lies farther away from a continental landmass than any other major volcanic islands in the world.

The plants and animals which colonized these Hawaiian Islands have developed into some of the most unique and specialized organisms anywhere. Every one of Hawaii's fifty-eight kinds of native forest birds are (or were) endemic to Hawaii, that is, they are found in no other location. Nearly 100 percent of all the insects and land snails in Hawaii are endemic, as are 95 percent of all flowering plants on the islands.

Tremendous changes have occurred in the islands' unique ecology since the first Polynesian settlers colonized the islands about 1,200 years ago. The Polynesians brought with them dogs, fowl, and about two dozen kinds of food and fiber plants, plus an unknown number of weeds and insects. They accidentally introduced rats to the islands. Large areas were cleared for agriculture, and over the years, fires set either intentionally or accidentally destroyed thousands of acres of dryland forests.

A second colonization of the islands began a little more than two centuries ago. Sea captains, seeking to provide their crews with a source of fresh meat, released cattle and sheep on Hawaii's lush plains. By 1825, the cattle and sheep, along with introduced horses, goats, and pigs, were multiplying rapidly in the absence of diseases and predators. Over the next century, these feral animals moved into the virgin forests, slowly destroying them and much of the bird life that lived there.

The rats accidentally brought to the islands by the Polynesians and later by Western colonists populated the islands. They no doubt are responsible for the extinction of some of the flightless birds on Laysan. The rats soon infested the great sugarcane plantations. In an effort to reduce the rat population, the Indian mongoose was introduced to the islands in 1883. The mongoose did not exterminate the rats, but it did become a serious predator on several of the native birds.

In more recent times, more than fifty species of birds and mammals have been introduced into the islands, including deer brought in for hunting purposes. Concurrently, native forest land was being cleared and replaced by foreign tree species thought to have greater potential as timber trees.

The decline of native kinds of animals and plants has been staggering. Of the seventy kinds of birds that were known to live on the Hawaiian Islands by the end of the eighteenth century, twenty-three have become extinct and thirty are on the current list of federally endangered species. When the United States Fish and Wildlife Service published their status review of threatened flowering plants on July 1, 1975, the list contained 639 potentially endangered plants, 193 potentially threatened plants, and 257 plants presumed already to be extinct.

The process of listing any of these endangered Hawaiian plants is a slow one. As of the end of August 1982, only five species from Hawaii had been listed as federally endangered.

Cooke's Kokio or *Kokia cookei.*

Mallow family. Shrub or small tree to 15 feet. Leaves alternate, 5- to 7-lobed. Flowers solitary in the axils of the leaves; petals 5, red, to 3 inches. Season: February to July.

Kokia cookei, or Cooke's kokio, a shrubby member of the mallow family that is closely related to cotton, has an interesting story of survival.

Sometime prior to 1871, three trees of this plant were found growing at the western end of the island of Molokai. The plants were about fifteen feet tall, and had gnarled trunks. Along the stem were several large, five- to seven-lobed leaves almost star-shaped in outline. Showy solitary red flowers were formed in the axils of some of the leaves.

When William Hillebrand published his *Flora of the Hawaiian Islands* in 1888, he had been unable to relocate this species. However, in April 1910 the Hawaiian plant collector Joseph Rock found one living and one dead tree of this species in a dry canyon at the west end of Molokai. It has been assumed that these were the same trees that were discovered four decades before.

Five years later, Rock and a rancher friend of his named George Cooke went back to the kokio site and found that the only living tree of this species was nearly dead. From that lone tree, however, Rock and Cooke collected a number of seeds. Several of these seeds germinated in a number of gardens, and several were scattered in wild areas on the island of Oahu. For some reason, most of the seedlings died, except for about thirty of them which Cooke was able to grow on his ranch. But the years took their toll on these garden plants. In 1971, years after Cooke's death, Keith Woolliams and Derral Herbst of the Pacific Tropical Botanical Garden on Kauai returned to Cooke's home and found only one plant alive. So for the second time, *Kokia cookei* was down to its last living specimen.

From this last tree, several seeds and cuttings were taken; several grafts were made. Few of these grew, except for one plant at the Waimea Arboretum on Oahu, five small grafted seedlings, and one seedling at the Royal Botanic Gardens in Kew, England.

On November 29, 1979, the United States Fish and Wildlife Service officially named *Kokia cookei* an endangered species.

Shrubby Sunflower or *Lipochaeta venosa*.

Aster family. Perennial; stems somewhat woody, sprawling or trailing, hairy. Leaves alternate, simple, hairy, blue-green, less than 2 inches. Flowers crowded together into a few heads with 6–8 yellow rays surrounding a central area of tubular flowers. Season: May, June.

In June 1910, while exploring the Nohonaohae Crater in the northwest quarter of the island of Hawaii, Joseph Rock, one of the finest plant collectors in the island's history, discovered a slightly woody, somewhat prostrate or trailing plant belonging to the sunflower family. The specimen remained unidentified until 1935, when Earl Edward Sherff of the Field Museum of Natural History in Chicago ran across it while doing some work on Hawaiian plants. Noting its differences from other plants in Hawaii, Sherff named the species *Lipochaeta venosa*, a shrubby sunflower known in Hawaii as nehe.

In 1938, E. Y. Hosaka made the second collection of this species near Rock's original find. It was growing in "dry places" at an altitude of about 3,000 feet. Hosaka, however, didn't realize he had found the nehe, as he identified the plant as something else. It wasn't until Robert Gardner was studying the genus *Lipochaeta* for his doctorate at Ohio State University in the 1970s that the true identity of Hosaka's plant became known.

Although Gardner, who completed his dissertation at Ohio State University in 1976, had not seen any specimens of *Lipochaeta venosa* that had been collected since 1949, his statement "probably extant" indicated he felt the plant still existed somewhere on the island of Hawaii.

Sure enough, when R. L. Stemmerman and C. H. Lamoureux were making a botanical survey as part of a team for the Army Corps of Engineers on the army's Pohakuloa Training Area, they found a sizable colony of nehe on the training grounds.

This colony, the only one known to be in existence for this species today, occurs in a rather dispersed area in ash-enriched soils on blocky lava outcroppings. It is found in habitats ranging from scrubby areas dominated by shrubs to scrub forests to *Euphorbia* or spurge forests.

Threats to this lone existing population include danger from heavy grazing by goats and sheep and disturbance by activities of the military, who use the area as a training ground. The United States Fish and Wildlife Service declared the species as endangered on October 30, 1979.

Lipochaeta venosa has harshly hairy stems that bear slender-stalked three-lobed leaves. The hairy, blue-green leaves, which are less than two inches long, are conspicuously veined on the lower surface. One or two flower heads occur at the end of each branchlet. Each head has six to eight yellow rays surrounding a central disk of twenty to thirty tubular flowers. The flowers bloom in May and June, usually following rain.

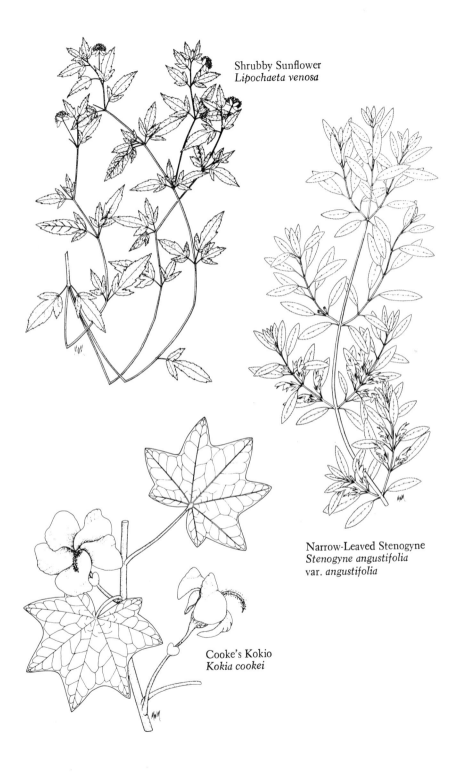

Shrubby Sunflower
Lipochaeta venosa

Narrow-Leaved Stenogyne
Stenogyne angustifolia
var. *angustifolia*

Cooke's Kokio
Kokia cookei

Hawaiian Wild Broad Bean or *Vicia menziesii*.
Pea family. High-climbing vine. Leaves alternate, compound, divided into 8, 10, or 12 toothless leaflets; tendrils present. Flowers in axillary, elongated clusters, pinkish, sweet-pea-shaped, to 1 inch. Season: June to October. (See color plate 40.)

Marvelous montane forests occur on the slopes of Mauna Loa and Mauna Kea on the island of Hawaii. A splendid variety of bird and plant life has always made these two beautiful mountains a paradise for naturalists.

Thus it was that Archibald Menzies made an expedition to Mauna Loa in 1794 to collect from the forested slopes of this mountain.

Many of the plants that Menzies found were new to science, including a handsome, rose-purple-flowered, high-climbing vetch of the pea family, which Menzies got from the forest edge at an altitude of 6,500 feet. When this vine was finally recognized as a new species, it was named *Vicia menziesii*. The common name that has developed for it is the Hawaiian wild broad bean.

In 1825, James Macrae collected this vine from the forests of Mauna Kea, a mountain adjacent to Mauna Loa. Sometime between 1851 and 1855, Jules Remy made a second collection of this species from Mauna Loa. It was the last time the Hawaiian wild broad bean was seen for nearly sixty-five years, until Charles Forbes of the Bishop Museum in Honolulu found twenty-four plants of *Vicia menziesii*. Then another drought of nearly sixty years elapsed, and the plant was thought to be extinct.

However, in 1974, while exploring the slopes of Mauna Loa, Wayne Gagné and Mae Mull discovered a small colony of the Hawaiian wild broad bean at an elevation of 5,200 feet. Shortly afterward, during the course of a forest bird study and later during a botanical survey of a 200-acre reforested area, two new small colonies of *Vicia menziesii* were found.

Although there is a possibility that this species may occur in other locations on Mauna Loa and Mauna Kea, it appears that this rare plant is found only in an area of a few hundred acres on Mauna Loa. At its present location, this species lives in a moist mountain forest dominated by two conspicuous trees of the Hawaiian forest, the koa (*Acacia koa*) and the ohia (*Metrosideros collina*). The habitat is transitional between the dense rain forest higher up the mountain slope and the more open, drier forests down the slope.

The decline of the Hawaiian wild broad bean probably began in the early 1800s, when cattle, sheep, goats, and pigs were introduced into the upland communities on the island of Hawaii. Not only do these animals graze upon the plants, they destroy much of the native vegetation, allowing alien elements to come into the area. These foreign species are aggressive and frequently "take over" an area.

Other factors which relate to the existence of this species are selective logging of the koas, which was begun in 1957, and the action of rodents. Derral Herbst of the United States Fish and Wildlife Service has observed plants of the Hawaiian wild broad bean whose main stem was chewed off at ground level by rodents.

With all of these perils, *Vicia menziesii* was designated a federally endangered species on April 26, 1978.

The Hawaiian wild broad bean is a perennial vine that climbs high on supporting vegetation. Each of its compound leaves is divided into eight, ten, or twelve toothless leaflets. A coiled tendril develops at the end of the leaf in place of a terminal leaflet. From the axils of some of the leaves is an elongated cluster of rather large, handsome, pinkish flowers.

Narrow-Leaved Stenogyne or
Stenogyne angustifolia var. *angustifolia.*
Mint family. Perennial; stems crawling or trailing. Leaves opposite, simple, leathery, to 2 inches, very narrow, finely toothed. Flowers to 1 inch, yellow to dull brownish-pink; petals 2-lipped. Season: Probably August to November.

Because of the remoteness of the Hawaiian Islands to a mainland continent, the plants and animals that are or were native to the islands have lived for a long time in isolation. Many groups of organisms have developed into clusters of different species, all unlike anything found on the mainland.

Several genera of plants have developed exclusively on the Hawaiian Islands. One of these plant groups is a genus of the mint family known as *Stenogyne*. Many stenogynes are either already extinct or are on their way to extinction. When the United States

Fish and Wildlife Service published its list of candidate species for possible endangered or threatened status on July 1, 1975, no fewer than eighteen kinds of stenogyne were listed as extinct, with thirteen more designated as potentially endangered and six as potentially threatened.

One of these, the narrow-leaved stenogyne, *Stenogyne angustifolia* var. *angustifolia*, has now been listed officially as a federally endangered plant on October 30, 1979.

Stenogyne angustifolia var. *angustifolia* is a narrow-leaved mint with stems that crawl or trail along the ground. The inch-long flowers are distinctly two-lipped, with the upper lip about twice as long as the lower one. The petals range from yellow to a dull brownish-pink.

The discovery of this plant was made on a United States South Pacific Exploring Expedition to the island of Hawaii in 1840 under the direction of Capt. Charles Wilkes of the United States Navy. About half a century later, William Hillebrand, who was working on his *Flora of the Hawaiian Islands*, collected this plant someplace on the islands, but Hillebrand failed to reveal the exact location.

Otto Degener and others in a collecting party on August 25, 1949, found the narrow-leaved stenogyne growing on the 1859 lava flow on the island of Hawaii, presumably near where they also found a specimen of the endangered shrubby sunflower (*Lipochaeta venosa*).

When Ray Fosberg and Derral Herbst published their list of rare and endangered species of Hawaiian vascular plants in 1975, they indicated that *Stenogyne angustifolia* var. *angustifolia* was probably extinct.

But on January 9, 1977, R. L. Stemmerman, working with a team conducting a botanical survey of the United States Army's Pohakuloa Training Area on the island of Hawaii, rediscovered this plant growing at an elevation of about 5,000 feet in an area transitional between a *Euphorbia* forest and a scrub forest community. The following year, a second colony was located on the Pohakuloa Training Area.

In 1977, one of these two populations, containing about ten plants, was destroyed by fire, leaving only a single colony of fifty or sixty plants. Although these plants seem to be holding their own, they are grazed upon occasionally by wild animals.

Haplostachys or *Haplostachys haplostachya* var. *angustifolia.*
Mint Family. Perennial; stems slightly woody, densely hairy, to
2 feet. Leaves opposite, simple, lance-shaped, velvety-hairy,
coarsely round-toothed, to 4 inches. Flowers many in an elon-
gated terminal spike, white with a purple tinge, to 1½ inches;
petals 2-lipped. Season: Probably August to November.

Haplostachys is a genus of mints composed of ten different kinds
of plants all restricted to the Hawaiian Islands. Because these
plants live in areas which have been depleted of their native vegetation during the past century, most of them were quickly wiped
out. When the United States Fish and Wildlife Service published
its list of potentially endangered and threatened species on July 1,
1975, it listed all ten kinds of haplostachys as extinct.

Discoveries of specimens in 1977 and 1978 of *Haplostachys
haplostachya* var. *angustifolia* at the United States Army's Pohakuloa Training Area have resurrected the genus.

There is an almost unbelievable parallelism among the collecting
history of the shrubby sunflower (*Lipochaeta venosa*), the narrow-
leaved stenogyne (*Stenogyne angustifolia* var. *angustifolia*), the
Hawaiian wild broad bean (*Vicia menziesii*), and the haplostachys,
all four federally endangered plants.

The first collection of haplostachys was made somewhere on the
island of Hawaii during the early 1850s by Jules Remy, the same
collector who found the Hawaiian wild broad bean on Mauna Loa.

In June 1910, Joseph Rock found the haplostachys on the No-
honaohae Crater, probably at the same time and place he found
the shrubby sunflower. The next year, Charles Forbes of the Bishop
Museum discovered a specimen of haplostachys on the slopes of
Mauna Kea. Four years later, Forbes was to find a colony of the
Hawaiian wild broad bean in the same area.

On August 25, 1949, Otto Degener and three others located the
shrubby sunflower, the narrow-leaved stenogyne, and the haplostachys all at the site of the 1859 lava flow. Then, during their
survey of the Pohakuloa Training Area in 1977 and 1978, Lamoureux, Stemmerman, and Warshauer found these same three kinds
of plants growing together again. All at one time were thought to
be extinct.

Haplostachys grows in full sun or partial shade in the *Euphorbia*
forest, where it lives on cinders or lava. There are several hundred

specimens scattered over several acres, but the number of plants apparently is decreasing. It was listed as endangered on October 30, 1979.

This mint is a slightly woody shrub that grows to a height of two feet. Many flowers are borne in an elongated, terminal spike. Each of the purplish-tinged white flowers is about one and one-half inches long and strongly two-lipped.

Ewa Plains Akoko or *Euphorbia skottsbergii* var. *kalaeloana*.

Spurge family. Upright shrub with milky sap. Leaves opposite, simple, oblong to elliptic, less than ½ inch. Flowers in small clusters in the leaf axils, inconspicuous, minute. Season: February to November.

On July 1, 1975, the United States Fish and Wildlife Service listed *Euphorbia skottsbergii* var. *kalaeloana*, a spurge of Hawaii's Ewa Plains, as extinct. This plant, along with a closely related variety, grew at one time on the Ewa Plains, an area of lowlands that extends from sea level to an elevation of about one hundred feet four to five miles inland along the coast of the island of Oahu at Barbers Point.

Because of great human disturbance throughout the years in the Ewa Plains, much of the natural vegetation has given way to introduced weedy plants. At one time, the plains were the site of cattle ranches, but now it is the location for the Barbers Point Naval Air Station, as well as a sugarcane plantation, an industrial park, some residential development, quarrying, and several pig and poultry farms. No wonder the two Euphorbias had become extinct. But wait! Derral Herbst, a botanist with the United States Fish and Wildlife Service, was observing the vegetation in the Ewa Plains in 1976 when he rediscovered some living plants of *Euphorbia skottsbergii* var. *kalaeloana*. The plants were growing on limestone where thickets of aggressive shrubs had taken over.

This plant had first been found by Charles Forbes of the Bishop Museum in 1912, and later by Joseph Rock, from the same area, in 1919. Herbst's rediscovery led to a survey of the plants of the Ewa Plains, conducted under the leadership of Wanda P. Char of Honolulu. During the survey, more than 5,000 plants of the Ewa Plains akoko were found at a half dozen different sites at Barbers Point.

Carter's Panic Grass
Panicum carteri

Haplostachys
Haplostachys haplostachya
var. *angustifolia*

Ewa Plains Akoko
Euphorbia skottsbergii var. *kalaeloana*

Since the area is still subject to further development, including the construction of a deep-draft harbor, the United States Fish and Wildlife Service proposed endangered status for this variety on September 2, 1980.

The Ewa Plains akoko is an upright shrub with many very slender branches. There are several pairs of opposite leaves on the stems. Most of the leaves are oblong to elliptic and have an asymmetrical base. They are less than one-half inch long. Inconspicuous clusters of minute flowers are formed in the axils of some of the leaves.

Carter's Panic Grass or *Panicum carteri*.

Grass family. Annual; stems slender, to 10 inches. Leaves very narrow, to 3 inches. Flowers inconspicuous; in short, dense clusters. Season: July to November.

The 132 islands that make up the Hawaiian chain stretch across 600 miles of Pacific Ocean. They range in size from islets of an acre or so to the island of Hawaii itself, which encompasses 4,038 square miles.

One tiny island that lies just off the eastern coast of Oahu at the head of Kaneohe Bay is Mokolii, sometimes known as Chinamans Hat. Mokolii, which barely covers four acres, is bordered by a number of rocky ledges.

Joseph Rock was the first to discover Carter's panic grass, but he did not know he had a new species when he found it near the seashore on Mokolii in 1917. In 1937, Raymond Fosberg collected the little grass on rock ledges on Mokolii, but he, too, failed to realize it to be a new species.

It wasn't until Edward Y. Hosaka of the Bishop Museum, collecting with Mitsugi Maneki on Mokolii on November 6, 1937, found the grass that it was recognized as a new species, which Hosaka called *Panicum carteri*.

In recounting his discovery, Hosaka noted that he "found twelve individual plants growing on a rocky ledge, which the salt spray drenches frequently."

Carter's panic grass occurs today in two restricted areas on Mokolii. Botanists in Hawaii fear that activities anywhere on the small island could be important to the survival of this species. A fire anywhere on the island during times of drought could conceivably destroy both populations of this grass. Even casual visitors to the islet could trample the few Carter's panic grasses that are left. It is feared that recent plantings of coconut palms on Mokolii may already have had an impact on this grass.

With these potential threats in mind, the United States Fish and Wildlife Service proposed early in 1981 that *Panicum carteri* be considered for endangered status. The proposal is pending at the time of this writing.

Panicum carteri is a delicate annual that grows in clumps up to about ten inches tall. The obscure grass flowers are .borne in very short, slender spikes that are barely exserted beyond the leaves.

Despite the fact that vast areas of wilderness in Alaska remain unexplored, there has been considerable botanical activity in the state. The first travelers to Alaska had other things on their minds besides plant collecting, but by 1900 the first botanists were making their way along the western coast. Later, as roads began penetrating the heartland of Alaska, collections from the interior were made.

Among the first of these early botanists was Jacob Peter Anderson. A native of Utah, Anderson received his botanical training at Iowa State University before going to Alaska in 1914. Over the years, Anderson made a great number of collections from Alaska, which culminated in the publication between 1943 and 1952 of his *Flora of Alaska and Adjacent Parts of Canada*. Several years later, Dr. Stanley Welsh was asked to update Anderson's flora. After studying all of Anderson's collections and making several expeditions to Alaska himself, Welsh published a new edition of *Anderson's Flora of Alaska* in 1974.

In the meantime, several persons living in Alaska became interested in the flora of that area, including a number of amateur botanists who organized the Juneau Botanical Club.

With the passage of the Endangered Species Act of 1973, interest increased in the rare plants of Alaska. David F. Murray of the Institute of Arctic Biology and University Museum of the University of Alaska at Fairbanks wrote a booklet on the *Threatened and Endangered Plants of Alaska* in 1980. Murray listed eight Alaskan plants which he thought to be endangered and thirteen he considered threatened. The federal government is

considering these for possible listings. Descriptions of four of the
rare Alaskan species follow.

Smelowskia or *Smelowskia pyriformis.*

*Mustard family. Tufted perennial; stems to 6 inches, with last
year's leaves persistent at base. Current year's leaves alternate,
simple, to 2 inches, 5- to 9-lobed. Flowers white, in terminal
clusters; sepals 4; petals 4, white, ⅛ inch. Pods pear-shaped.
Season: July, August.*

One of these rare Alaskan plants is a small member of the mustard
family known as *Smelowskia pyriformis.* This plant was unknown
to science until 1949, when W. H. Drury, Jr., who was studying
the genus *Smelowskia* at Harvard University, made a collecting
trip to Alaska and discovered a plant of this new species growing
in limestone on top of Farewell Mountain. This mountain is
located just a little south of the center of the state. A few years
later, a second population of the smelowskia was discovered on a
rocky limestone slope in a nearby area. Drury and Reed C. Rollins,
also of Harvard University, named this new plant *Smelowskia
pyriformis.*

The stems of this perennial are only six inches long, at most.
Terminating each stem are branched clusters of many small, white
flowers. Each of the four petals of the flower is only about one-
eighth inch long, but these are longer than the four green sepals.
There are also six stamens within the flower. The fruit that develops from each flower is a small, pear-shaped pod. The flowers
bloom during the summer.

Kobuk Locoweed or *Oxytropis kobukensis.*

*Pea family. Perennial; stems to 6 inches, with last year's leaf
stalks persistent at base. Leaves alternate, compound, with 13–17
lance-shaped leaflets to ½ inch. Flowers purple, ½ inch long,
on a leafless stalk; petals sweet-pea-shaped. Season: June, July.*

Up in the far northwestern corner of Alaska flows the Kobuk
River, a scenic and interesting clear-flowing stream. Near the junction of the Kobuk and Hunt rivers, for a stretch of about twenty-
five miles, is a unique area of sand dunes which have piled up
along the Kobuk River.

This little-known section of Alaska was the subject of a plant-

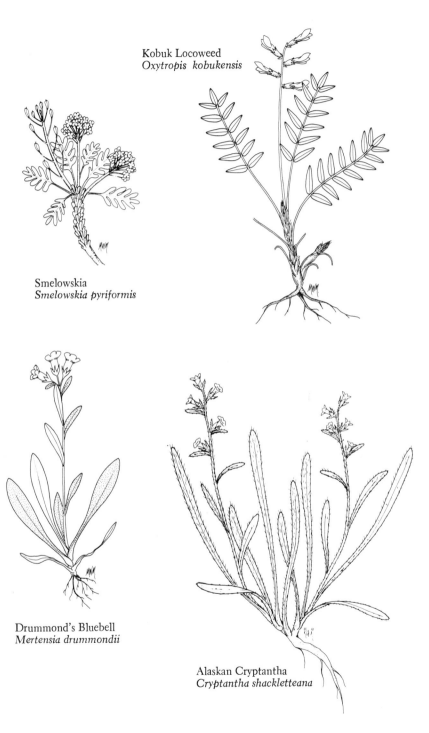

Kobuk Locoweed
Oxytropis kobukensis

Smelowskia
Smelowskia pyriformis

Drummond's Bluebell
Mertensia drummondii

Alaskan Cryptantha
Cryptantha shackletteana

collecting trip made by R. D. Hamilton during July 1938. While examining the vegetation growing in the sand dunes along the Kobuk River, Hamilton found a small member of the pea family, but failed to recognize it as a new species. But when Dr. Stanley Welsh of Brigham Young University, who was studying both the flora of Alaska and the genus *Oxytropis*, ran across Hamilton's collection, he realized it was unlike any other *Oxytropis* anywhere in the world. Welsh named it *Oxytropis kobukensis*.

Although Welsh recognizes a dozen or so different kinds of *Oxytropis* in Alaska, *Oxytropis kobukensis* is the only one that grows in sand dunes, although some of the others have an affinity for sandy soils.

The Kobuk locoweed is a perennial herb up to six inches tall. The lower part of each branch is very distinctive in that it is densely covered with old persistent leaf stalks that are reddish to purple-brown.

Each compound leaf is divided into thirteen to seventeen leaflets which are lance-shaped and about one-half inch long. There are five or six flowers borne at the end of a leafless stalk that arises from the base of the plant. Each of the purplish, sweet-pea-shaped flowers is about two-thirds of an inch long.

Drummond's Bluebell or *Mertensia drummondii*.

Borage family. Perennial; stems to 8 inches, smooth. Basal leaves narrow, to 4 inches long; leaves on stem smaller. Flowers several at the tip of the stem, blue, tubular, to ⅔ inch long. Season: July, August.

Wildflower enthusiasts are usually ecstatic over finding patches of plants known as bluebells. In the eastern United States, the Virginia bluebell (*Mertensia virginica*) brings expressions of approval from nature lovers, while in the West, the tall chiming bells (*Mertensia ciliata*) draw equal raves from the western counterparts. Bluebells and chiming bells belong to the genus *Mertensia*, a group of some twenty-five North American plants and additional ones in Asia. Three of these are found in Alaska, with the rarest and most restricted of them being Drummond's bluebell, *Mertensia drummondii*.

This bluebell is confined in Alaska to a series of sand dunes along the Meade River in the most northern region of Alaska. The area is part of the Arctic tundra north of the Brooks Range. This species also occurs to a limited extent in the western Canadian Arctic from Victoria Island and the Dolphin and Union straits.

Mertensia drummondii is a perennial that grows to a height of about eight inches. Several blue, tubular flowers are borne at the tip of the stem. The flowers bloom during July and August.

Alaskan Cryptantha or *Cryptantha shackletteana*.

Borage family. Annual; stems to 6 inches, very hairy. Leaves very narrow, to 4 inches long, to ¼ inch broad, hairy. Flowers few to several near the tip of the stem, white, to ¼ inch long. Season: July, August.

The genus *Cryptantha* is a common group of wildflowers scattered in much of the western United States. Many of them are known as nievitas; one is called the miner's candles. Still others are simply referred to as cryptanthas. In Alaska, however, there is only one native kind of cryptantha, and it is one of the rarest plants in the state. Known as *Cryptantha shackletteana*, this rather small, densely hairy plant has been found only twice, both in the vicinity of Eagle in the upper Yukon River region near the Canadian border.

Where it was originally found, the Alaskan cryptantha was growing on an extremely steep, dry, rocky slope. More recently, a second colony was discovered at the margin of sparse grasslands. Both populations are many miles from the nearest kind of cryptantha.

Cryptantha shackletteana is an annual that reaches a height of about six inches. A number of small white flowers are formed near the tip of the stem during July and August.

It's Up to Us

W here have all the wildflowers gone? Some, like *Thismia*, have become extinct and are gone forever. Others, such as some of the 120 plants discussed in the preceding chapters, are only a step or two away from extinction.

Thismia is not alone as an extinct plant. Many others have not been seen for years and are presumed to be extinct. That's too bad, because I would like to have seen *Hechtia texensis*, a prickly member of the pineapple family that was found on dry limestone bluffs in the Big Bend area of Texas in 1885 and not seen since.

Wouldn't it be nice to go to the Undive Falls in Yellowstone National Park and see the showy rock cress, *Arabis fructicosa?* Nobody has been able to do so since 1899. Or go to Klamath Falls in Oregon and see Applegate's milk vetch, *Astragalus applegatii?* It was found in 1927 and again in 1931, but not since.

Porter's goldenrod, *Solidago porteri*, was found early this century near Monticello, Georgia, and in Jackson County, North Carolina, but botanists have searched high and low for it in recent years without success. Smith's sunflower, *Helianthus smithii*, was found in the mountains of Rabun County, Georgia, and Randolph County, Alabama, in 1884, but no more.

It may not have seemed like much when the little annual San Francisco allocarya (*Plagiobothrys diffusa*) disappeared from San Francisco in 1933, but it was an irreversible loss. Somehow it should have been kept alive. I would have liked that. Or how about *Congdonia pinetorum*, a sedumlike succulent that was last seen on the eastern slopes of the Sierra Nevadas near Mammoth Lakes, California, a few years ago? Surely it won't be missed, for there are a lot of sedumlike succulents still around to take its place. Not quite. Its genetic makeup, its potential, perhaps as an ornamental—

or whatever else the Creator had in mind for it—will never be realized. I miss it, because it was alive at one time, and perhaps it could have been saved had any of us cared.

The ashy phacelia (*Phacelia cinerea*) from San Nicholas Island, California, is gone, and so is the stipulate scurf pea (*Psoralea stipulata*), last seen in the wild in Indiana in 1856, and the Ocmulgee skullcap (*Scutellaria ocmulgee*), a mint that was found for the last time near Macon, Georgia, in 1898.

Wouldn't you like to have seen the spurless small-flowered columbine, *Aquilegia micrantha* var. *mancosana*, growing along the Mancos River in Colorado? Alice Eastwood saw it in 1892. Lucky for her.

The Roan Mountain goat's-beard, *Astilbe crenatiloba*, hasn't been seen this century along the North Carolina–Tennessee border, nor has Whipple's monkeyflower, *Mimulus whipplei*, from rocky places in a pine forest in Calaveras County, California.

Some extinctions are recent, when some of us alive today perhaps could still have done something about it if only we had cared about endangered species before the 1970s.

The Marin County paintbrush (*Castilleja leschkeana*) survived at Point Reyes until 1960. The checker mallow (*Sidalcea keckii*) lasted in Tulare County, California, until 1939. Oregon's Sexton Mountain mariposa lily persisted until at least 1948. Unfortunately, the list goes on and on.

We have taken steps to stem the flow toward extinction. The Endangered Species Act of 1973 and its subsequent amendments and reauthorizations have given us the foundation to work with.

A few foresighted persons saw the plight many years ago. In the United States, Dr. William Trelease, director of the Missouri Botanical Garden at the turn of the century, in his address to the American Association for the Advancement of Science at their New York meeting in June 1900, appealed for the protection of our native plants and the organization of local societies for that purpose. This may have prompted the organization in Boston later that year of the Society for the Protection of Native Plants.

Later, the Wild Flower Preservation Society became an incorporated organization in Washington, D. C. Even during the mid-1930s, this society ran the following notice in every issue of their journal *Wild Flower*:

The disappearance of 99.99% of all wild flowers has been and will continue to be due, not to picking, but to over grazing, fires and deforestation with the resulting erosion and siltage; also to agricultural, commercial and real estate developments.

The cutting of Dogwood trees for spindles; the stripping of bark from the trunks of roots of Dogwood, Redbud and Black Haw, and the digging of roots of Ginseng, Goldenseal, Lobelia, Ladyslipper, Trillium, Lily-of-the-valley, Bloodroot and many others, which are of doubtful value, for the drug trade; the wasteful pulling of Trailing Arbutus runners; the commercial digging of the roots of Orchids, Gentians, Lilies and many other attractive rare flowers for gardens, and the collecting of Christmas Greens, Galax leaves for decorating and Deertongue for Tobacco flavoring, is causing the rapid disappearance of these and other plants from many localities. Unless these are soon brought into cultivation, most of those employed in the trade will have to seek some other source of income.

Concentrate on publicity campaigns with the cooperation of the press; extend the cultivation of rare species, but only from seeds, cuttings and root divisions unless they can be obtained from land that is to be cleared; establish wild flower preserves; urge nurserymen to do more propagating of rare species; urge the setting aside of restricted and protected areas in all public parks and forests; urge the increase of the native species in restricted park and forest areas and the establishment of nature trails in them.

If these statements were relevant in 1930, they are more so today. Every one of us can do something about it. Many native plant societies have been organized in the last few years. These are organizations dedicated to the preservation of our wild flora.

Several national organizations are making concentrated efforts to assist endangered species. The Nature Conservancy buys critical parcels of land in order to save endangered species. The Garden Club of America has begun its "Adopt a Flower" program. In this program, each participating club of the 14,500-member Garden Club of America has "adopted" a plant that is endangered or threatened in its area (although not necessarily on the federal list). Club members are currently sending postcards or photographs of their adopted flowers to the United States Fish and Wildlife Service as part of their campaign to dramatize the necessity of protecting the rare, threatened, or endangered flowers in their areas.

So get involved. Join an organization. Take botanical field trips and be thankful that the plants you see are still alive. Who knows, you may discover that *Thismia americana* is alive and well somewhere. It's up to us.

Appendix 1

The Federal Government's List of Endangered and Threatened Plants

Scientific Name	Common Name	Status	Date Added to List	States
Aconitum novebora-cense	Northern Monkshood	T	4–26–78	IA, NY, OH, WI
Ancistrocactus tobuschii	Tobusch's Fishhook Cactus	E	11–7–79	TX
Arabis macdonaldiana	McDonald's Rock Cress	E	9–28–78	CA, OR
Arctomecon humilis	Dwarf Bear Poppy	E	11–6–79	UT
Arctostaphylos hookeri ssp. ravenii	Raven's Man-zanita	E	10–26–79	CA
Astragalus perianus	Rydberg's Milk Vetch	T	4–26–78	UT
Baptisia arachnifera	Hairy Rattle-weed	E	4–26–78	GA
Berberis sonnei	Sonne's Bar-berry	E	11–6–79	CA
Betula uber	Virginia Round-Leaf Birch	E	4–26–78	VA
Callirhoe scabriuscula	Texas Poppy Mallow	E	1–13–81	TX
Castilleja grisea	San Clemente Island Paintbrush	E	8–11–77	CA
Cordylanthus mariti-mus ssp. maritimus	Salt Marsh Bird's Beak	E	9–28–78	CA

Scientific Name	Common Name	Status	Date Added to List	States
Coryphantha minima	Nellie Cory Cactus	E	11-7-79	TX
Coryphantha ramillosa	Bunched Cory Cactus	T	11-6-79	TX
Coryphantha sneedii var. leei	Lee's Pincushion Cactus	T	10-25-79	NM
Coryphantha sneedii var. sneedii	Sneed's Pincushion Cactus	E	11-7-79	TX, NM
Delphinium kinkiense	San Clemente Island Larkspur	E	8-11-77	CA
Dudleya traskiae	Santa Barbara Island Live-Forever	E	4-26-78	CA
Echinacea tennesseensis	Tennessee Coneflower	E	6-6-79	TN
Echinocactus horizonthalonius var. nicholii	Nichol's Devil's Head Cactus	E	10-26-79	AZ
Echinocereus engelmannii var. purpureus	Engelmann's Purple Hedgehog Cactus	E	10-11-79	UT
Echinocereus kuenzleri	Kuenzler's Hedgehog Cactus	E	10-26-79	AZ
Echinocereus lloydii	Lloyd's Hedgehog Cactus	E	10-26-79	NM, TX
Echinocereus reichenbachii var. albertii	Albert's Black Lace Cactus	E	10-26-79	TX
Echinocereus triglochidiatus var. arizonicus	Arizona Red Claret Cactus	E	10-25-79	AZ
Echinocereus triglochidiatus var. inermis	Spineless Red Claret Cactus	E	11-7-79	CO, UT
Echinocereus viridiflorus var. davisii	Davis' Green Pitaya	E	11-7-79	TX
Eriogonum gypsophilum	Gypsum Wild Buckwheat	T	1-19-81	NM

Scientific Name	Common Name	Status	Date Added to List	States
Erysimum capitatum var. angustatum	Contra Costa Wallflower	E	4–26–78	CA
Haplostachys haplostachya var. angustifolia	Haplostachys	E	10–30–79	HA
Harperocallis flava	Harper's Beauty	E	10–2–79	FL
Hedeoma apiculatum	McKittrick Pennyroyal	T	7–13–82	NM, TX
Hedeoma todsenii	Todsen's Pennyroyal	E	1–19–81	NM
Hudsonia montana	Mountain Golden Heather	T	10–20–80	NC
Isotria medeoloides	Small Whorled Pogonia	E	9–10–82	CT, IL, MA, MD, ME, MI, MO, NC, NH, NJ, NY, PA, RI, SC, VA, VT
Kokia cookei	Cooke's Kokio	E	11–29–79	HA
Lipochaeta venosa	Shrubby Sunflower	E	10–30–79	HA
Lotus dendroideus ssp. traskiae	San Clemente Island Broom	E	8–11–77	CA
Malacothamnus clementinus	San Clemente Island Bushmallow	E	8–11–77	CA
Mirabilis macfarlanei	MacFarlane's Four-O'clock	E	10–26–79	OR, ID
Neolloydia mariposensis	Lloyd's Mariposa Cactus	T	11–6–79	TX
Oenothera avita ssp. eurekensis	Eureka Dunes Evening Primrose	E	4–26–78	CA
Oenothera deltoides ssp. howellii	Antioch Dunes Evening Primrose	E	4–26–78	CA

Scientific Name	Common Name	Status	Date Added to List	States
Orcuttia mucronata	Pointed Orcutt Grass	E	9–28–78	CA
Pedicularis furbishiae	Furbish's Lousewort	E	4–26–76	ME
Pediocactus bradyi	Brady Cactus	E	10–26–79	AZ
Pediocactus knowltonii	Knowlton's Cactus	E	10–26–79	NM
Pediocactus peeblesi- *anus*	Peebles Navajo Cactus	E	10–26–79	AZ
Pediocactus sileri	Siler Pin-cushion Cactus	E	10–26–79	AZ, UT
Phacelia argillacea	Clay Phacelia	E	9–28–78	UT
Pogogyne abramsii	San Diego Mesa Mint	E	9–28–78	CA
Potentilla robbinsiana	Dwarf Cinquefoil	E	9–17–80	NH, VT
Rhododendron chap- *manii*	Chapman's Rhododen-dron	E	4–24–79	FL
Sagittaria fasciculata	Bunched Arrowhead	E	7–25–79	NC, SC
Sarracenia oreophila	Green Pitcher Plant	E	12–19–79	AL, GA
Sclerocactus glaucus	Uinta Hook-less Cactus	T	10–11–79	CO, UT
Sclerocactus *mesae-verdae*	Mesa Verde Cactus	E	10–30–79	CO, NM
Sclerocactus wrightiae	Wright's Fishhook Cactus	E	10–11–79	UT
Spiranthes parksii	Navasota Ladies'-Tresses Orchid	E	5–6–82	TX
Stenogyne angustifolia var. *angustifolia*	Narrow-Leaved Stenogyne	E	10–30–79	HA

Scientific Name	Common Name	Status	Date Added to List	States
Swallenia alexandrae	Eureka Dunegrass	E	4–26–78	CA
Trillium persistens	Persistent Trillium	E	4–26–78	GA, SC
Vicia menziesii	Hawaiian Wild Broad Bean	E	4–26–78	GA
Zizania texana	Texas Wild Rice	E	4–26–78	CA

Appendix 2
Status of Other Species
in This Book

The status of the remaining fifty-six species discussed in this book is: C—candidate for Federal endangered listing as of September 1, 1982; N—nominated plants for state endangered or threatened listing; SC—plants of special concern to states but which are not currently being reviewed by Federal government.

Scientific Name	Common Name	Status	States
Aeschynomene virginica	Sensitive Joint-Vetch	N	DE, MD, NC, NJ, PA, VA
Ammobroma sonorae	Sand Food	N	AZ, CA
Apios priceana	Price's Groundnut	N	AL, IL, KY, MS, TE
Asclepias meadii	Mead's Milkweed	N	IA, IL, IN, KA, MO, WI
Aster depauperatus	Depauperate Aster	N	DE, MD, PA
Astragalus detritalis	Debris Milk Vetch	SC	CO, UT
Astragalus montii	Heliotrope Milk Vetch	C	UT
Astragalus yoder-williamsii	Osgood Mountain Milk Vetch	N	ID, NV
Bidens bidentoides	Tidal Shore Beggar's-Tick	N	DE, MD, NJ, NY, PA
Cereus robinii	Florida Tree Cacti	N	FL
Chrysosplenium iowense	Iowa Golden Saxifrage	N	IA, MN
Cirsium pitcheri	Great Lakes Thistle	N	IL, IN, MI, WI
Conradina verticillata	Cumberland Rose-mary	N	KY, TE
Coryphantha recurvata	Golden-Chested Beehive Cactus	N	AZ

Scientific Name	Common Name	Status	States
Cryptantha shackletteana	Alaskan Cryptantha	N	AK
Cypripedium arietinum	Ram's-Head Lady's-Slipper Orchid	SC	CT, MA, ME, MI, MN, NH, NY, VT, WI
Darlingtonia californica	Cobra Plant	N	CA, OR
Dionaea muscipula	Venus Fly Trap	N	FL, NC, SC
Dodecatheon frenchii	French's Shooting Star	N	AR, IL, IN, KY, MO
Draba aprica	Whitlow Grass	N	AR, GA, MO, OK, SC
Eriocaulon kornickianum	Pipewort	N	AR, OK, TX
Euphorbia purpurea	Purple-Flowered Spurge	N	DE, MD, NC, PA, VA, WV
Euphorbia skottsbergii var. *kalaeloana*	Ewa Plains Akoko	C	HI
Eutrema penlandii	Colorado Eutrema	N	CO
Ferocactus acanthodes var. *eastwoodiae*	Yellow-Spined Barrel Cactus	N	AZ
Geocarpon minimum	Geocarpon	N	AR, MO
Geum radiatum	Appalachian Avens	N	NC, TE
Glaucocarpum suffrutescens	Rollins' Thelypody	N	UT
Helianthemum dumosum	Frostweed	N	CT, MA, NY, RI
Heterotheca ruthii	Ruth's Golden Aster	N	TE
Iliamna remota	Kankakee Mallow	N	IL, VA
Iris lacustris	Lake Iris	N	MI, WI
Lilium parryi	Lemon Lily	N	AZ, CA
Lobelia laxiflora var. *angustifolia*	Red Lobelia	SC	AZ
Mertensia drummondii	Drummond's Bluebell	N	AK
Oxytropis kobukensis	Kobuk Locoweed	N	AK
Panicum carteri	Carter's Panic Grass	C	HI
Paronychia argyrocoma var. *albimontana*	Silverling	C	MA, ME, NH
Penstemon grahamii	Graham's Beardstongue	N	CO, UT
Petalostemum foliosum	Leafy Purple Prairie Clover	N	AL, IL, TE
Phacelia formosula	Handsome Phacelia	C	CO
Plantago cordata	Heart-Leaved Plantain	N	AL, GA, IL, IN, MI, MO, NC, NY, OH, WI

Scientific Name	Common Name	Status	States
Platanthera integra	Southern Yellow Orchid	N	AL, FL, GA, LA, MS, NC, NJ, SC, TE, TX
Prenanthes boottii	Boott's White Lettuce	N	ME, NH, NY, VT
Pseudotaenidia montana	Mountain Pimpernel	N	MD, PA, VA, WV
Sarracenia alabamensis	Canebrake Pitcher Plant	N	AL
Shortia galacifolia	Oconee Bells	N	GA, NC, SC
Sida hermaphrodita	Virginia Mallow	SC	MD, OH, TE, VA, WV
Smelowskia pyriformis	Smelowskia	⁻N	AK
Solidago houghtonii	Houghton's Goldenrod	N	MI, NY
Solidago spithamaea	Blue Ridge Goldenrod	N	NC
Sullivantia sullivantii	Sullivant's Sullivantia	N	IN, KY, OH
Synandra hispidula	Synandra	N	AL, IL, IN, KY, NC, TN, VA, WV
Trifolium virginicum	Kates Mountain Clover	N	MD, PA, VA, WV
Trollius laxus	Spreading Globe Flower	N	CT, NJ, NY, OH, PA
Viola novae-angliae	New England Violet	N	ME, MI, MN, NY, WI

Appendix 3

Plants Not Found in Several Years and Presumed to Be Extinct

Plant	Family	Last Seen	State
Agalinis caddoensis	Snapdragon	1913	LA
Agalinis stenophylla	Snapdragon	1897	FL
Amsinckia carinata	Borage	1896	OR
Aquilegia micrantha var. mancosana	Buttercup	1892	CO
Arabis fructicosa	Mustard	1899	WY
Arctostaphylos hookeri ssp. franciscana	Heath	1968	CA
Arenaria franklinii var. thompsonii	Carnation	1955	OR
Astilbe crenatiloba	Saxifrage	1895	NC, TE
Astragalus acnicidus	Pea	1954	CA
Astragalus applegatii	Pea	1931	OR
Astragalus desereticus	Pea	1909	UT
Astragalus humillimus	Pea	1875	CO
Astragalus kentrophyta var. douglasii	Pea	1883	OR, WA
Astragalus pycnostachyus var. lanosissimus	Pea	1967	CA
Atriplex tularensis	Goosefoot	1921	CA
Bacopa simulans	Snapdragon	1941	VA
Calamagrostis nubila	Grass	1862	NH
Calochortus indecorus	Lily	1948	OR
Calochortus monanthus	Lily	1876	CA
Calyptridium pulchellum	Portulaca	1938	CA
Campanula robbinsiae	Bellflower	1958	FL
Carex livida	Sedge	1922	CA
Carex specuicola	Sedge	1948	AZ
Castilleja leschkeana	Snapdragon	1960	CA

Plant	Family	Last Seen	State
Castilleja ludoviciana	Snapdragon	1915	LA
Chorizanthe valida	Buckwheat	1967	CA
Clarkia mosquinii ssp. xerophila	Evening Primrose	1968	CA
Congdonia pinetorum	Sedum	1939	CA
Cryptantha aperta	Borage	1892	CO
Cryptantha insolita	Borage	1905	NV
Cuscuta warneri	Dodder	1957	UT
Dissanthelium californicum	Grass	1912	CA
Echinocereus hempelii	Cactus	1897	NM
Elodea linearis	Waterweed	1875	TE
Elodea nevadensis	Waterweed	1887	NV
Eriophyllum lanatus	Aster	1905	CA
Eriophyllum latilobum	Aster	1956	CA
Eriophyllum nubigenum	Aster	1883	CA
Franklinia altamaha	Tea	1790	GA
Gerardia acuta	Snapdragon	1928	NY
Hechtia texensis	Pineapple	1885	TX
Hedeoma pilosum	Mint	1940	TX
Helianthus nuttallii ssp. parishii	Aster	1937	CA
Helianthus praetermissus	Aster	1929	NM
Helianthus smithii	Aster	1884	AL, GA
Hemizonia mohavensis	Aster	1933	CA
Howellia aquatilis	Bellflower	1928	CA
Ivesia callida	Rose	1923	CA
Juncus pervetus	Rush	1928	MA
Lathyrus hitchcockianus	Pea	1893	CA
Lechea mensalis	Rockrose	1931	TX
Limosella pubiflora	Carrot	1928	AZ
Lomatium nelsonianum	Carrot	1946	OR
Lupinus cusickii var. abortivus	Pea	1896	OR
Lycium hassei	Nightshade	1936	CA
Lycium verrucosum	Nightshade	1901	CA
Malacothamnus abbottii	Mallow	1889	CA
Malacothamnus mendocinensis	Mallow	1938	CA
Matelea texensis	Milkweed	Not known	TX
Micranthemum micranthemoides	Snapdragon	1941	VA
Mimulus brandegei	Snapdragon	1932	CA
Mimulus pygmaeus	Snapdragon	1894	CA
Mimulus traskiae	Snapdragon	1904	CA
Mimulus whipplei	Snapdragon	1854	CA
Monardella leucocephala	Mint	1941	CA
Monardella pringlei	Mint	1921	CA
Nephropetalum pringlei	Chocolate	1888	TX
Opuntia strigil var. flexospina	Cactus	1911	TX
Orthocarpus pachystachyus	Snapdragon	1913	CA

Plant	Family	Last Seen	State
Pedicularis crenulata	Snapdragon	1933	CA
Perityle villosa	Aster	1935	CA
Phacelia amabilis	Waterleaf	1942	CA
Phacelia cinerea	Waterleaf	1901	CA
Phacelia nevadensis	Waterleaf	1867	NV
Phacelia pallida	Waterleaf	1883	WA
Phacelia parishii	Waterleaf	1941	CA
Physaria grahamii	Mustard	1930	UT
Pisonia floridana	Four-o'clock	1870	FL
Plagiobothrys diffusus	Borage	1933	CA
Plagiobothrys hirtus ssp. coralli-carpa	Borage	1961	OR
Plagiobothrys lamprocarpus	Borage	1921	OR
Pleuropogon oregonus	Grass	1937	OR
Polygonum montereyense	Buckwheat	1917	CA
Potentilla multijuga	Rosa	1890	CA
Psoralea macrophylla	Pea	1897	NC
Psoralea stipulata	Pea	1871	IN
Ranunculus acriformis var. aestivalis	Buttercup	1945	CA
Rorippa coloradensis	Mustard	1875	CO
Salvia blodgettii	Mint	Prior to 1860	FL
Scheuchzeria palustris var. americana	Arrow-grass	1897	CA
Scutellaria ocmulgee	Mint	1898	GA
Sesuvium trianthemoides	Carpetweed	1947	TX
Seymeria havardii	Snapdragon	1882	TX
Sibara filifolia	Mustard	1901	CA
Sidalcea keckii	Mallow	1939	CA
Silene rectiramea	Carnation	1935	AZ
Solanum bahamense var. rugelii	Nightshade	1860	FL
Solanum carolinense var. hirsutum	Nightshade	1860s	GA
Solidago porteri	Aster	1899	NC
Suaeda duripes	Goosefoot	1860s	TX
Synthyris missurica ssp. hirsuta	Snapdragon	1881	OR
Thismia americana	Burmannia	1913	IL
Trisetum orthochaetum	Grass	1908	MT

Excluded from this list are more than 250 Hawaiian plants that are presumed to be extinct.

Appendix 4

Glossary

areole. An area on a cactus stem from which arise the spines.

awn. A very slender terminal projection.

axil. The junction where a leaf is attached to a stem.

basal leaves. Leaves that are confined to the base of the plant, at ground level.

bract. A structure that subtends a flower, often looking like a small leaf.

catkin. An elongated cluster of flowers that lacks petals.

cherty. Referring to rocks that are flintlike and usually contain iron.

cleft. Deeply notched.

corolla. All the petals of a flower considered together.

lanceolate. Shaped like a lance, tapering from the base to the tip.

lip. One part of the petals of a member of the mint or snapdragon families. Each flower usually has an upper lip and a lower lip. In the orchid family, the lip is a highly modified petal.

lobe. A rounded projection along the edge of a leaf.

marly. Referring to soil that is a mixture of clay and sand.

ovate. Egg-shaped but flattened, not having depth.

palmately. Divided in a manner resembling the fingers of a hand.

pistil. That part of the flower that produces the immature seeds.

pistillate. Referring to a flower that has pistils but no stamens.

ray. A petallike structure found in most members of the aster family.

rhizome. A much-branched stem found underground.

ribs. Narrow and elongated ridges on a cactus stem, usually oriented vertically or diagonally.

sepal. The outermost, usually green, part of a flower.

sessile. Without a stalk.

sheath. A transparent structure that usually serves as protection for whatever it surrounds.

spine. A hard, pointed projection.

stamen. That part of the flower that produces the pollen.

staminate. Referring to a flower that has stamens but no pistils.

stipule. A leaflike structure found at the point of attachment of the leaf to the stem in some plants.

stolon. A horizontal stem lying on the surface of the soil.

succulent. Any plant structure plump with stored water.

tendril. A coiled structure that enables some vines to climb.

umbel. A cluster of flowers with the stalks arranged like the spokes of an umbrella.

Index